Natural Products and Derivatives in Human Disorders

Natural Products and Derivatives in Human Disorders

Editors

Eva E. Rufino-Palomares
José Antonio Lupiáñez

MDPI • Basel • Beijing • Wuhan • Barcelona • Belgrade • Manchester • Tokyo • Cluj • Tianjin

Editors
Eva E. Rufino-Palomares
Biochemistry and Molecular Biology I
University of Granada
Granada
Spain

José Antonio Lupiáñez
Biochemistry and Molecular Biology I
University of Granada
Granada
Spain

Editorial Office
MDPI
St. Alban-Anlage 66
4052 Basel, Switzerland

This is a reprint of articles from the Special Issue published online in the open access journal *Molecules* (ISSN 1420-3049) (available at: www.mdpi.com/journal/molecules/special_issues/natural_products_chemoprevention).

For citation purposes, cite each article independently as indicated on the article page online and as indicated below:

LastName, A.A.; LastName, B.B.; LastName, C.C. Article Title. *Journal Name* **Year**, *Volume Number*, Page Range.

ISBN 978-3-0365-2767-3 (Hbk)
ISBN 978-3-0365-2766-6 (PDF)

Contents

About the Editors

Eva E. Rufino-Palomares

Eva E. Rufino-Palomares graduated in Biology from the University of Granada (UGR). In 2005, she obtained a pre-doctoral scholarship in public competition with which she obtained a PhD degree in 2009. Later, she had a research contract (2009-2011) and, finally, she obtained the position of Assistant Professor in the Department of Biochemistry and Molecular Biology I of the UGR (2011). Currently, she is Senior Lecturer. She has a total of 35 scientific contributions. She has been on the research team in 12 long-term research projects and 3 as principal. She presents important contributions. She presents a university teaching experience of 14 years. She has directed 3 doctoral, 20 final master projects, and 14 final degree projects, and has been responsible for 11 research initiation scholarships for undergraduate and postgraduate students. She has directed two teaching innovation projects, and she is secretary of the academic commission of the biochemistry degree.

José Antonio Lupiáñez

José Antonio Lupiáñez was born in Almería (Spain). He studied Biology at the University of Granada between 1967 and 1975, obtaining a BSc degree in 1972, a MSc degree in 1973 and a PhD degree in 1975. In 1973, he spent a semester as a predoctoral fellow at Cardiff University. In 1978, he obtained a postdoctoral Fulbright Fellowship in the Pharmacology department of the Indiana University School of Medicine under the supervision of professor S.R. Wagle. Upon his return to Spain, after two years, he obtained a position as Associate Professor and founded the research group "Drugs, environmental toxics and cellular metabolism". In 1990, he received the position of Full Professor. He has directed a score of doctoral theses, and is the author of over 200 publications. He is a member of different scientific societies, the Spanish Society of Biochemistry and Molecular Biology, European Society of Biochemistry, New York Academy of Sciences and American Chemical Society. In October 2018, he was appointed as Permanent Professor Emeritus.

Preface to "Natural Products and Derivatives in Human Disorders"

When the *Molecules* Editorial Office proposed to us the idea of coordinating the edition, as Academic Editors, of a Special Issue related to our research experience carried out during the last twenty-five years, we were tremendously attracted to the possibility of being able to make a monographic book related to the bioactive role of different natural compounds from the kingdoms of plants and fungi, as well as many of their chemical derivatives, in the solution of a wide group of human disorders and pathologies. After deep reflection and after rejecting several titles, we decided to title it "Natural Products and Derivatives in Human Disorders".

This Special Issue was finished forging with the intention of requesting, collecting and disseminating some of the most significant and recent contributions in the use of natural compounds and derivatives to reduce the risk of developing inflammatory and oxidant diseases such as cancer and others human disorders.

Natural products are bioactive compounds synthesized by terrestrial and marine plants, fungi, microorganisms and animals. Traditionally, they have been used in the prevention and treatment of various human diseases in different cultures. In parallel, chemical derivatives of these natural compounds have been used in order to enhance their bioactivities.

Although numerous nitrogenous compounds present in the composition of all living organisms such as amino acids, nucleotides, biogenic amines, purine and pyrimidine bases and many others play an important role in their own development and survival [1,2], it is an authentic reality that scientists have recognized, for more than 50 years, that the wealth of secondary products also recognized as phytochemicals in the plant kingdom in secondary products, is much greater when compared to those found in the animal kingdom.

As we noted before [3,4], at the dawn of the 19th century, numerous researchers isolated and characterized, from a chemical point of view, numerous secondary products in different groups of plants that they called phenols, isoprenoids and alkaloids capable of presenting a real physiological and biochemical activity whose initial purpose was that of defense against different types of external aggressions caused by herbivores, competitors, pathogens and predators, avoiding, on the one hand, recurrent infections caused by parasites, bacteria and viruses, and on the other hand, the destruction of the organism produced by different types of phytophagous and mycophagous [5]. Later, from the 40s and 50s of the previous century, and with the advance of analytical techniques and methodologies, more than 5000 secondary products could be recognized and characterized, a number that nowadays exceeds, by far, 100,000 compounds [6].

During the last ten years, most of them have been reported to have a variety of interesting and significant biological properties, such as analgesic, anti-allodynic, anti-diabetic, anti-oxidant, anti-parasitic, antimicrobial, anti-viral, anti-atherogenic, anti-inflammatory, anti-proliferative, anti-tumour, growth-stimulating activities, as well as cardio and neuro-protective activities [7-11]. However, special attention has been focused on the study of their anti-tumour capacity, through the potential modulation of cancer initiation and growth, cellular differentiation, apoptosis and autophagy, angiogenesis, and metastatic dissemination. Moreover, a considerable number of studies reported the relation of the anticancer effect with their anti-inflammatory and antioxidant activities. Besides these capacities, the use of natural compounds and derivatives represent one of the most promising strategies to treat metabolic disorders such as oxidative stress, diabetes, obesity, metabolic syndrome, and so on [12-16].

The topics included in this Special Issue are:

·New natural compounds and derivatives as anticancer agents.

·Use of natural compounds and derivatives as anti-inflammatory agents in human disorders.

·Natural products and derivatives in oxidative stress associated with human diseases.

·Molecular mechanism implicated of natural products and derivatives.

·Effects of natural products and derivatives on nutrition and diet on human diseases.

·Type of natural product and derivatives with potential bioactive, extraction and synthesis.

After a relatively long period of requests for contributions, we decided to choose, from a total of more than 30 papers submitted, the seven contributions that passed the peer review and in which more than twenty independent reviewers participated. Finally, this book compiles a total of one editorial and seven original articles in which different phytochemicals and chemical derivatives are used, delving into the study of the molecular roles they present in different human pathologies, such as cancer, neuronal and ocular dysfunctions, arterial hypertension and osteoarthritis.

The different contributions presented in this book have been made by 55 authors of up to six nationalities (Korea, Spain, Germany, China, United Arab Emirates and Oman). The topics covered include situations that alter pathological growth processes of the liver, colon and breasts, with situations related to neurodegenerative diseases through inhibitors of acetyl and butyrylcholinesterases, and with alterations of the vascular, bone and ocular systems such as hypertension, osteoarthritis and dry eye disease, also known as keratoconjunctivitis.

The academic editors of this work thank the MDPI editorial group for this scientific space that it offers us and that serves to share with its future readers the feelings generated by the compilation of the different works that make up this work. We also want to congratulate all the participating authors, all the reviewers who without their work make it difficult for science to continue advancing and all the scientific and administrative staff of the *Molecules* Editorial Office for their extraordinary and excellent work. Finally, our special wish is that we all enjoy your reading.

References

1. de Paz-Lugo, P.; Lupiáñez, J.A.; Meléndez-Hevia, E. High glycine concentration increases collagen synthesis by articular chondrocytes in vitro - Acute glycine deficiency could be an important cause of osteoarthritis. Amino Acids 2018, 50, 1357-1365. Doi: 0.1007/s00726-018-2611-x.

2. de Paz-Lugo, P.; Meléndez-Hevia, E. Branch-point stoichiometry can generate weak links in metabolism: the case of glycine biosynthesis. J. Biosci. 2008; 33, 771-780. Doi: 10.1007/s12038-008-0097-5.

3. Lupiáñez, J.A.; Rufino-Palomares, E.E. Phytochemicals: "A Small Defensive Advantage for Plants and Fungi; A Great Remedy for the Health of Mankind". Molecules 2021, 26, 6159. Doi: 10.3390/molecules26206159.

4. Lupiáñez, J.A.; Rufino-Palomares, E.E.; Pérez-Jiménez, A. (Eds.). Anticancer Properties of Natural and Derivative Products, 1st ed.; ©1996-2021 MDPI: Basel, Switzerland, 2021, pp. 1-311.

5. Molyneux, R. J.; Lee, S.T.; Gardner, D.R.; Panter, K.E.; James, L.F. Phytochemicals: The good, the bad and the ugly? Phytochemistry 2007, 68, 2973-2985. Doi: 10.1016/j.phytochem.2007.09.004.

6. Harborne, J. B.: Arsenal for survival: secondary plant products. Taxon 2000, 49, 435-449. Doi: 10.2307/1224343.

7. Rufino-Palomares, E.E.; Reyes-Zurita, F.J.; Fuentes-Almagro, C.A; de la Higuera, M.; Lupiáñez, J.A.; Peragón, J. Proteomics in liver of Gilthead Sea bream (Sparus aurata) to elucidate the cellular response induced by the intake of maslinic acid. Proteomics 2011, 11, 3312-3325. Doi: 10.1002/pmic.201000271.

8. Reyes-Zurita, F.J.; Rufino-Palomares, E.E.; Medina, P.P.; García-Salguero, L.; Peragón, J.; Cascante, M.; Lupiáñez, J.A. Antitumour activity on extrinsic apoptotic targets of the triterpenoid maslinic acid in p53-deficient Caco-2 adenocarcinoma cells. Biochimie 2013, 95, 2157–2167. Doi: 10.1016/j.biochi.2013.08.017.

9. Mokhtari, K.; Rufino-Palomares, E.E.; Pérez-Jiménez, A.; Reyes-Zurita, F.J.; Figuera, C.; García-Salguero, L.; Medina, P.P.; Peragón, J.; Lupiáñez, J.A. Maslinic Acid, a Triterpene from Olive, Affects the Antioxidant and Mitochondrial Status of B16F10 Melanoma Cells Grown under Stressful Conditions. Evid. Based Complement. Altern. Med. 2015, 2015, 1–11. Doi: 10.1155/2015/272457.

10. Reyes-Zurita, F.J.; Rufino-Palomares, E.E.; García-Salguero, L.; Peragón, J.; Medina, P.P.; Parra, A.; Cascante, M.; Lupiáñez, J.A. Maslinic Acid, a Natural Triterpene, Induces a Death Receptor-Mediated Apoptotic Mechanism in Caco-2 p53-Deficient Colon Adenocarcinoma Cells. PLoS ONE 2016, 11, e0146178. Doi: 10.1371/journal.pone.0146178.

11. Pérez-Jiménez, A.; Rufino-Palomares, E.E.; Fernández-Gallego, N.; Ortuño-Costela, M.C.; Reyes-Zurita, F.J.; Peragón, J.; García-Salguero, L.; Mokhtari, K.; Medina, P.P.; Lupiáñez, J.A. Target molecules in 3T3-L1 adipocytes differentiation are regulated by maslinic acid, a natural triterpene from Olea europaea. Phytomedicine 2016, 23, 1301–1311. Doi: 10.1016/j.phymed.2016.07.001.

12. Parra, A.; Martin-Fonseca, S.; Rivas, F.; Reyes-Zurita, F.J.; Medina-O'Donnell, M.; Martinez, A.; García-Granados, A.; Lupiáñez, J.A.; Albericio, F. Semi-synthesis of acylated triterpenes from olive-oil industry wastes for the development of anticancer and anti-HIV agents. Eur. J. Med. Chem. 2014, 74, 278–301. Doi: 10.1016/j.ejmech.2013.12.049.

13. Parra, A.; Martin-Fonseca, S.; Rivas, F.; Reyes-Zurita, F.J.; Medina-O'Donnell, M.; Rufino-Palomares, E.E.; Martínez, A.; Garcia-Granados, A.; Lupiáñez, J.A.; Albericio, F. Solid-Phase Library Synthesis of Bi-Functional Derivatives of Oleanolic and Maslinic Acids and Their Cytotoxicity on Three Cancer Cell Lines. ACS Comb. Sci. 2014, 16, 428–447. Doi: 10.1021/co500051z.

14. Reyes-Zurita, F.J.; Medina-O'Donnell, M.; Ferrer-Martin, R.M.; Rufino-Palomares, E.E.; Martin-Fonseca, S.; Rivas, F.; Martínez, A.; García-Granados, A.; Pérez-Jiménez, A.; García-Salguero, L.; et al. The oleanolic acid derivative, 3-O-succinyl-28-O-benzyl oleanolate, induces apoptosis in B16–F10 melanoma cells via the mitochondrial apoptotic pathway. RSC Adv. 2016, 6, 93590–93601. Doi: 10.1039/c6ra18879f.

15. Medina-O'Donnell, M.; Rivas, F.; Reyes-Zurita, F.J.; Martinez, A.; Martin-Fonseca, S.; Garcia-Granados, A.; Ferrer-Martín, R.M.; Lupiáñez, J.A.; Parra, A. Semi-synthesis and antiproliferative evaluation of PEGylated pentacyclic triterpenes. Eur. J. Med. Chem. 2016, 118, 64–78. Doi: 10.1016/j.ejmech.2016.04.016.

16. Medina-O'Donnell, M.; Rivas, F.; Reyes-Zurita, F.J.; Martinez, A.; Lupiáñez, J.A.; Parra, A. Diamine and PEGylated-diamine conjugates of triterpenic acids as potential anticancer agents. Eur. J. Med. Chem. 2018, 148, 325–336. Doi: 10.1016/j.ejmech.2018.02.044.

Eva E. Rufino-Palomares, José Antonio Lupiáñez
Editors

molecules

MDPI

Editorial

Phytochemicals: "A Small Defensive Advantage for Plants and Fungi; A Great Remedy for the Health of Mankind"

José A. Lupiáñez * and Eva E. Rufino-Palomares *

Department of Biochemistry and Molecular Biology I, Faculty of Sciences, University of Granada, Avenida Fuentenueva, 1, 18071 Granada, Spain

* Correspondence: jlcara@ugr.es (J.A.L.); evaevae@ugr.es (E.E.R.-P.); Tel.: +34-958-243089 (J.A.L.); +34-958-243252 (E.E.R.-P.); Fax: +34-958-249945 (J.A.L. & E.E.R.-P.)

check for updates

Citation: Lupiáñez, J.A.; Rufino-Palomares, E.E. Phytochemicals: "A Small Defensive Advantage for Plants and Fungi; A Great Remedy for the Health of Mankind". *Molecules* **2021**, *26*, 6159. https://doi.org/10.3390/molecules26206159

Received: 17 September 2021
Accepted: 11 October 2021
Published: 12 October 2021

In the chronology of Biochemistry, as a new science that emerged in the mid-nineteenth century after its separation from Organic Chemistry and Physiology, its beginnings were characterized by an intense search and subsequent isolation and characterization of different organic compounds that were part of the chemical composition of living organisms. Scientists, such as Schwann, Pasteur, Berthelot, Bernard, Liebig, Wöhler and Büchner, played a fundamental role in these origins. For example, Schwann discovered in 1836 that gastric fluid contained, in addition to hydrochloric acid, another digestive component that he called pepsin, or Wöhler, who obtained in 1828 an organic molecule, urea, using the method of chemical synthesis that bears his name. All these initial studies formed what, for a long time, was known as Elemental and Structural Biochemistry. The compilation of all these organic elements for more than a hundred years served as the chemical basis for the classification of most of the natural compounds present in the biosphere [1].

The next step in the "construction" of Biochemistry, well into the 20th century, was an intensive study of the functionality and dynamics of each of the organic molecules present in living beings. Some of them are used as ashlars in the construction of macromolecules destined, mainly, in the formation of the macrostructures of the living being, while others play an effective functional and dynamic role. In this way, the circle of the structure-function binomial, basic for the understanding of all biological sciences, is closed. All this constitutes what is known as Functional and Dynamic Biochemistry, and although the first known vestiges of metabolism date back to the middle of the 13th century by the hand of a doctor from Damascus called Ibn al-Nafis (1213–1288), who stated, in his best known work, *Theologus Autodidactus*, that "... the body and all its parts are in a continuous state of dissolution and nutrition, so they are inevitably in permanent change", its greatest exponent was HA Krebs, Nobel Prize in Physiology or Medicine in 1953, after the publication in 1932 of his work on the discovery of the urea cycle, a cyclical metabolic process through which different nitrogenous compounds, mainly amino acids, are processed when they are not recycled, generating urea as a final product [2].

From a functional point of view, a good example of the latter class of molecules is a group of natural compounds known as "phytochemicals." An extraordinary, and almost inexhaustible source of these compounds is made up of a large part of terrestrial and aquatic plants, along with numerous species from the kingdom of fungi. Over time, it has become known that these molecules play a decisive role in the defence of these plants and fungi against herbivores, competitors, pathogens and predators, preventing, on the one hand, recurrent infections caused by parasites, bacteria and viruses, and on the other, of the destruction of the organism produced by different types of phytophages and mycophages [3]. A large number of these compounds are found in foods and although they are not considered as nutrients or macronutrients, nor are they included in the group of vitamins or minerals, they provide various beneficial functions. That is why foods that contain them are called functional foods since, in addition to their nutritional component, they also provide other types of health benefits.

In view of how the development of phytochemicals has evolved, and paraphrasing the recapitulation theory of Ernst Haeckel, who in 1866 defended that " ... embryonic ontogeny is a brief and rapid recapitulation of phylogeny", it can be said, in a certain way, that the history of phytochemicals recapitulates the history of biochemistry; first it was the discovery of molecules together with the decipherment of their structure and later, the recognition of their functions and bioactivities. At present, and especially in the last 20 years, there is a real explosion in the search for these compounds, capable of presenting important and abundant biological properties. Although many of them act as real poisons and teratogens, many more are being investigated thanks to their beneficial effects on health and can be used as new drugs or adjuvants in the treatment of various diseases [4]. Taking advantage of the anti-inflammatory capacities of many of them, their therapeutic use could be very useful in all those infectious diseases with a huge increase in cytokines, as is the case of the pandemic produced by SARS-CoV-2. In this sense, in several countries a clinical trial is being carried out using maslinic acid (pentacyclic triterpene) together with hydroxytyrosol (polyphenol), which have both demonstrated high antiviral and anti-inflammatory capabilities, to treat COVID-19 patients.

The classification of organic compounds has always been very ambiguous and that is also the case of phytochemicals. However, five groups can be included among them as the most significant: flavonoids, phytosterols, terpenoids, lignans and stilbenes [5], all of which are recognized for their large number of bioactive effects as nutraceuticals, essential nutrients and even allelopathic, thus influencing both the growth and survival, as well as the reproduction of other organisms [6]. Among all these phytochemicals, polyphenols and triterpenoids should be highlighted, as many of their bioactive capacities are known, both in vitro and in vivo [4]. Many of these compounds have various beneficial properties, such as anticancer, antiproliferative and antiangiogenic in different tumour lines [7–15], as well as antioxidant and anti-inflammatory [16–19], while others have properties related to metabolic syndromes, such as antidiabetogenic and cardio and neuroprotective [20–22], others anti-infectious, such as antifungal, antimicrobial, antiviral and antiparasitic [23–25], and, finally, others that affect organic and metabolic activity, such as growth inducers, activators and inhibitors of enzymatic activity [26–30], as well as modulators in the production of reducing equivalents in the form of NADPH. This explains most biosynthetic, cellular and organic growth, nutritional processes, as well as processes of differentiation, cellular detoxification and oxygen free radical scavenging [31–37].

On the other hand, the process of organic synthesis has been used recently with increasing intensity, with the objective of finding chemical derivatives of natural compounds that present improvements in functional effectiveness in relation to the bioactivity of the original compound. In this sense, different chemical groups are being used in the synthesis of these derivatives, significantly increasing their effectiveness [4]. Among these groups, acyls, aminoacyls and dipeptidyls [23,24,38,39], derivatives of pegylated and diamino-pegylated groups [40–44], and even derivatives with coumarin [45], with which their activities are significantly increased, stand out.

After a rigorous peer review process, the Special Issue entitled Natural Products and Derivatives in Human Disorders, (https://www.mdpi.com/journal/molecules/special_issues/natural_products_chemoprevention, date accessed 20 January 2019), compiles a total of seven original articles in which different phytochemicals and chemical derivatives are used, delving into the study of the molecular roles they present in different human pathologies, such as cancer, neuronal and ocular dysfunctions, arterial hypertension and osteoarthritis. A summary of the main characteristics of the original papers included in this Special Issue, both in terms of provenance and molecular type, as well as their molecular effects, are discussed below.

It is relatively frequent that pathologies related to the production of tears by the lacrimal glands can occur both in elderly people and in patients with certain types of autoimmune diseases or even by external aggressions such as chemical or thermal burns. When this problem worsens, the well-known dry eye disease (DED) or dry keratoconjunctivitis appears, the main pathology of which consists of corneal ulcers and infections. The most common treatment is to moisten the eye with artificial tears composed basically with hyaluronic acid, or with lubricating ointments; Kang et al. [46], in a very interesting study, shows the beneficial effects of the extract of *Aucuba japonica* or spotted laurel, together with its bioactive compound, aucubin, an iridoid glycoside, that is, a monoterpene glycosylated derivative. In their study, these authors demonstrate that in in vitro assays, human corneal cells (PCS-700-010), exposed to desiccation stress and treated with these compounds, extract and active ingredient, increase their survival in a clearly dose-dependent effect. At the same time, overexpression of mRNA levels, generated by ocular desiccation, of different inflammatory cytokines (interleukins 1β (IL-1β) and 8 (IL-8), and tumor necrosis factor TNF-α) were significantly reduced by the administration of the extract and aucubin. Moreover, in in vivo tests using SD (Sprague-Dawley) rats as an animal model, they found that after unilateral excision of the exorbital lacrimal gland, both lacrimal volumes and corneal irregularities recovered completely. In addition, they found that these compounds significantly reduced apoptotic cells in the cornea. All these results strongly suggest that these compounds can be considered as a novel therapy for this disease and that aucubin is probably responsible for this effect.

The evolutionary appearance of vertebrate animals generated the need to implement an efficient bone turnover system that allows them to carry out an adequate regeneration of the entire osteoarticular system. To this end, it is essential to ensure a regulated and constant synthesis of collagen. A key element for this is a constant and adequate supply of glycine molecules [47,48]. However, the cellular synthesis of this amino acid is greatly reduced as it is linked to the concomitant synthesis of coenzymatic forms of tetrahydrofolic acid (THF), preventing the production of sufficient and adequate amounts of this amino acid, thus fulfilling the restriction theorem of glycine biosynthesis which states: "*If the only significant metabolic pathway for glycine synthesis is the reaction catalyzed by glycine hydroxymethyl transferase, the steady-state metabolic flux for net glycine production cannot, under any condition, exceed the flux for the consumption of the C1 units transferred* via N^5-N^{10} *methylene THF*". Glycine must therefore be considered an essential amino acid because the capacity for its synthesis is much lower than its actual requirement and, as a consequence, the inevitable onset of degenerative diseases such as osteoarthrosis, osteoporosis and osteoarthritis [47,48]. Therefore, all these bone diseases are inevitably linked to the growth and development of large vertebrates. Within metabolic theory, this evolutionary "failure", especially frequent and serious in large vertebrates, is classified as a so-called "default metabolic error". In this context, traditional Korean medicine has been using an herbal preparation known as RyuPungHwan (RPH) consisting of extracts of seven different plants, *Astragalus membranaceus, Turnera diffusa, Achyranthes bidentata, Angelica gigas, Eclipta prostrata, Eucommia ulmoides* and *Ilex araguariensis*, with high content of flavonoids and triterpenoids, in patients with osteoarthritis, not to cure the disease but to alleviate the inflammatory process and pain. Taking into account this background, Hong et al. [49], have studied the effects of this herbal preparation in SW1353 chondrosarcoma cell models, taking into account the inflammatory process associated with the disease. Administration of this preparation produced a significant inhibition of Interleukin-1β-stimulated inflammation, making two active compounds isolated from the herbal preparation, isomucronulatol 7-O-β-D-glucoside, a flavonoid derivative, and ecliptasaponin A, a triterpenoid derivative, responsible for this clear anti-inflammatory effect. They showed that both the preparation and the two identified compounds were able to reduce the expression of matrix metalloproteinase 13 (MMP13), cyclooxygenases 1 and 2 (COX-1 and -2), Tumor necrosis factor α (TNF-α), Interleukin-1β (IL-1β) or protein p65, previously increased by the administration of IL-1β to SW1353 cells.

Hypertension, which is very common nowadays, is a serious risk factor associated with cardiovascular diseases, which, together with other diseases such as obesity, diabetes and atherosclerosis, constitute the characteristic tetrad of the metabolic syndrome. In this context, Hong et al. [50], belonging to the same research group as the previous work, analyze the effects on the main molecular markers of hypertension of a plant preparation, commonly used in traditional Korean medicine and consisting of three species, *Pine densiflora*, *Annona muricate* and *Momordica charantia*, called No-ap (NA), as well as several of its main bioactive molecules, a roseoside (a compound belonging to the group of C13-norisoprenoids generated by the degradation of 40-carbon terpenes) and the flavonoid, icariside E4. The authors used as a biological model H9C2 cells, rat cardiomyocytes derived from myoblasts, treated with angiotensin II as a hypertensive molecule; with an antihypertensive drug, termisartan; with gingenoside as a positive control; with different doses of the NA extract; and with the two isolated compounds (roseoside and icariside), both individually and together. Treatment with angiotensin II resulted in a significant increase in myocyte angiotensin II receptor 1, (AT1), tumor necrosis factor α, (TNF-α), monocyte chemoattractant protein 1, (MCP-1), tumor growth factor β, (TGF-β) mRNA expression levels, NADPH oxidase enzyme activity, H_2O_2 and superoxide ion ($\bullet O_2^-$) levels, while the activity levels of the antioxidant enzymes catalase and superoxide dismutase were significantly reduced. Subsequent treatment with the extract and the isolated components restored the levels of all markers to control values. The effect found was dose-dependent and, in addition, a synergistic effect of the active compounds was observed. The authors conclude that hypertension therapy with these compounds should be considered for further clinical trials.

Neurodegenerative diseases constitute, nowadays, a serious problem, afflicting people of all ages, but especially older people, preventing them from adequate cognitive development and thus seriously affecting their life and daily activities. Many are classified as neurodegenerative diseases and among the most important and visible are Alzheimer's disease and dementia with Lewy bodies, as well as Friedreich's ataxia, Huntington's disease and Parkinson's disease. In the first two, but especially in Alzheimer's disease, one of the most important biochemical dysfunctions is the appearance of very high activity of the neurotransmitter degradation systems, mainly acetylcholine. Therefore, the role of cholinesterases, especially acetylcholinesterase and butyrylcholinesterase, plays a central role in the development of this disease. It is therefore of interest to develop potential drugs, based on natural compounds, that are capable of slowing down the rapid elimination of these neurotransmitters. In this regard, Loesche et al. [51] have succeeded in synthesizing up to 40 types of carboxamides, molecules derived from ethylenediamine, from up to five different naturally occurring triterpenes, oleanolic, ursolic, maslinic, betulinic and platanic acids. These authors evaluated acetylcholinesterase and butyrylcholinesterase activities using the Ellman assay and both enzymes were derived, respectively, from *Electrophorus electricus* and equine serum. Among all the carboxamides tested, seven different compounds showed inhibition constant (Ki) values in the nanomolar to micromolar range. The inhibitions are of mixed type with competitive dominance, according to the molecular modelling studies performed. Therefore, the synthesis of molecules with these characteristics could be used in the therapy of this type of neurodegenerative disease.

Teucrium mascatense is an understory shrub, the genus of which is widely distributed over several continents. Its medicinal value has been known since ancient times and, therefore, it has been used in both traditional and modern medicine, thanks to the presence of its bioactive components, including tannins, glycosides, phenols, steroids and terpenoids. Its bioactivities include antibacterial, antipyretic, anti-inflammatory, antidiabetic, antioxidant, analgesic and antipyretic properties. In this case, Panicker et al. [52] have investigated the anticancer capacity of extracts from this plant, as well as that of one of its main components, triterpene IM60. In their studies they used four cell lines, one for cervical cancer, HeLa and three for breast cancer, MCF-7, MDA-MB-231 and MCF-10A. To certify the possible antiproliferative activity, they studied the effects of the extract and the isolated triterpene on some of the most important apoptotic markers, such as cytotoxicity at different times and concentrations; levels of caspases 3, 7, 8 and 9; levels of PARP (Poly ADP ribose polymerase); measurements of cell apoptosis by Annexin V/Propidium iodide assays in order to decipher the different types of apoptosis, early, late and necrosis, as well as the morphological changes associated with apoptosis. The results indicate a clear anticancer effect by significantly increasing the levels of all molecular markers together with an increase in the apoptosis of malignant cells. This study identifies the triterpene IM60 as a potential drug that can be successfully used in breast cancer therapy.

If it is important that certain compounds have the ability to cure diseases, some of them as serious as cancer, in which irreparable individual and collective problems are caused, the ability to prevent them is an incalculable advance, both from the personal point of view and social, by being able to avoid all its fatal consequences. We know that many of the phytochemicals present in nature have this property and we just need to be able to find them and find it. This is the case of the good work presented by Juan et al. [53]. In it, the chemopreventive capacity of maslinic acid, isolated from *Olea europaea*, is studied in vivo in an animal model that mimics sporadic human colorectal cancer. Male Sprague-Dawley rats were used in this test, to which for 49 days they were administered, orally, different doses of maslinic acid, 5, 10 or 25 mg kg of animal weight. Subsequently, and after a week of rest, the cancer was induced by means of three weekly injections of 20 mg/kg of the carcinogenic inducer, 1,2-dimethylhydrazine. Under these conditions, different preneoplastic markers, aberrant crypt foci (ACF) and mucin-depleted foci (MDF) were analyzed. Since the administration of the lowest dose of maslinic acid, 5 mg/kg, decreases of 15% were found in ACF and up to 27% in MDF, achieving these significant decreases with the 25 mg/kg dose, with those that the reductions of these biomarkers were 33% and 50%, respectively, in addition, these results were corroborated by their association with the concentrations of maslinic acid found in the colon of the animals. The results of this work clearly demonstrate the preventive role that this phytochemical has on colon cancer, which allows us to conclude that with an adequate addition of this triterpene in the diet, this chemopreventive activity could be achieved, so important as to eliminate the problems inherent of this disease.

The inclusion of bioinformatics studies in biological research is essential to advance our understanding of the molecular mechanisms involved in the functions being sought. At present, a very high percentage of molecular biology studies use this essential tool [10]. The work presented by Lv et al. [54] falls within this experimental concept, in which experimental studies are combined with computational studies. In their work, the authors use a saponin (glycosylated triterpene), saikosaponin D, from a preparation of Chinese origin known as Radix Bupleuri, used as a traditional medicine for over 2,000 years in China, Japan, Korea and other Asian countries, and consisting of the dried roots of two plant species, *Bupleurum chinense* and *Bupleurum scorzonerifolium*. It has previously been reported that saikosaponin D is able to inhibit cancer cell growth by inducing apoptosis and arresting the cell cycle in the G_1 phase. The present work aims to further investigate the molecular anticancer mechanism of this saponin by combining both network pharmacology and metabolomics databases. Up to 35 targets were studied in the bioinformatics analysis, selecting neuropilin-1 (NRP-1) for further investigation based on the degree of molecular docking score, demonstrating that saikosaponin D combined with NRP-1 deletion could significantly ameliorate HepG2 damage. As a consequence, metabolomics analyses showed that NRP-1 blockade exhibited the lowest metabolite deregulation score, and among the metabolites analyzed, carnitines and short- or long-chain phospholipids were mainly implicated. These results clearly implicate NRP-1 as a key molecule to explain the anti-hepatoma activity of saikosaponin D.

An example of the importance the use of different phytochemicals has had in recent years in order to help prevent, alleviate or, in its best version, cure the numerous pathologies that currently afflict humanity, is precisely the content of this Special Issue that we are commenting on. Diseases such as dry keratoconjunctivitis, osteoarthrosis, hypertension as a risk factor in cardiovascular diseases, the role of certain cholinesterase inhibitors in some neurodegenerative diseases, or some types of cancers, such as breast, colon and liver, are the subject of the works presented here. For all these reasons, the academic editors of this issue would like to sincerely thank all our colleagues, authors and reviewers, whose efforts and collaboration have contributed to the realization of this Special Issue, allowing it to become a reality.

Author Contributions: Study design, J.A.L.; Data Collect, J.A.L.; Data analysis and interpretation, J.A.L. and E.E.R.-P.; Writing—original draft preparation, J.A.L.; Writing—review and editing. J.A.L. and E.E.R.-P. All authors have read and agreed to the published version of the manuscript.

Funding: This work did not receive external funding.

Institutional Review Board Statement: Not applicable.

Informed Consent Statement: Not applicable.

Data Availability Statement: Not applicable.

Acknowledgments: The authors thank Alejandro L. Lupiáñez for the critical revision of the English version of the text. We are also grateful to all colleagues, authors and reviewers whose efforts have contributed to the realization of this special issue. And of course, we would like to thank the MDPI publishing group and the Molecules editorial team for their help, constant support and professionalism.

Conflicts of Interest: The authors of this work declare that its content was carried out in the absence of any commercial or financial relationship and, therefore, does not present any conflict of interest now or in the future.

References

1. Cordón, F. *History of Biochemistry. Historical-Critical Consideration of Biochemistry from the Theory of Biological Levels of Integration*, 1st ed.; Literary Company L.P.: Madrid, Spain, 1997; pp. 1–319. ISBN 84-8213-038-2.
2. Krebs, H.A.; Henseleit, K. Untersuchungen über die Harnstoffbildung im tierkorper. *Z. Physiol. Chem.* **1932**, *210*, 33–66. [CrossRef]
3. Molyneux, R.J.; Lee, S.T.; Gardner, D.R.; Panter, K.E.; James, L.F. Phytochemicals: The good, the bad and the ugly? *Phytochemistry* **2007**, *68*, 2973–2985. [CrossRef] [PubMed]
4. Rufino-Palomares, E.E.; Pérez-Jiménez, A.; Reyes-Zurita, F.J.; García-Salguero, L.; Mokhtari, K.; Herrera-Merchán, A.; Medina, P.P.; Peragón, J.; Lupiáñez, J.A. Anti-cancer and Anti-angiogenic Properties of Various Natural Pentacyclic Tri-terpenoids and Some of their Chemical Derivatives. *Curr. Org. Chem.* **2015**, *19*, 919–947. [CrossRef]
5. Dinkova-Kostova, A. Phytochemicals as Protectors against Ultraviolet Radiation: Versatility of Effects and Mechanisms. *Planta Med.* **2008**, *74*, 1548–1559. [CrossRef] [PubMed]
6. Lupiáñez, J.A.; Rufino-Palomares, E.E.; Pérez-Jiménez, A. Are Ancestral Medical Practices the Future Solution to Today's Medical Problems? *Molecules* **2021**, *26*, 4701. [CrossRef] [PubMed]
7. Reyes-Zurita, F.J.; Centelles, J.J.; Lupiáñez, J.A.; Cascante, M. (2α, 3β)-2,3-Dihydroxyolean-12-en-28-oic acid, a new natural triterpene from *Olea europea*, induces caspase dependent apoptosis selectively in colon adenocarcinoma cells. *FEBS Lett.* **2006**, *580*, 6302–6310. [CrossRef]
8. Laszczyk, M.N. Pentacyclic Triterpenes of the Lupane, Oleanane and Ursane Group as Tools in Cancer Therapy. *Planta Med.* **2009**, *75*, 1549–1560. [CrossRef]
9. Rufino-Palomares, E.E.; Reyes-Zurita, F.J.; García-Salguero, L.; Mokhtari, K.; Medina, P.P.; Lupiáñez, J.A.; Peragón, J. Maslinic acid, a triterpenic anti-tumoural agent, interferes with cytoskeleton protein expression in HT29 human colon-cancer cells. *J. Proteom.* **2013**, *83*, 15–25. [CrossRef]
10. Sánchez-Tena, S.; Reyes-Zurita, F.J.; Diaz-Moralli, S.; Vinardell, M.P.; Reed, M.; Garcia-Garcia, F.; Dopazo, J.; Lupiáñez, J.A.; Günther, U.; Cascante, M. Maslinic Acid-Enriched Diet Decreases Intestinal Tumorigenesis in ApcMin/+ Mice through Transcriptomic and Metabolomic Reprogramming. *PLoS ONE* **2013**, *8*, e59392. [CrossRef]
11. Reyes-Zurita, F.J.; Rufino-Palomares, E.E.; Medina, P.P.; García-Salguero, L.; Peragón, J.; Cascante, M.; Lupiáñez, J.A. Antitumour activity on extrinsic apoptotic targets of the triterpenoid maslinic acid in p53-deficient Caco-2 adenocarcinoma cells. *Biochimie* **2013**, *95*, 2157–2167. [CrossRef]
12. Reyes-Zurita, F.J.; Rufino-Palomares, E.E.; García-Salguero, L.; Peragón, J.; Medina, P.P.; Parra, A.; Cascante, M.; Lupiáñez, J.A. Maslinic Acid, a Natural Triterpene, Induces a Death Receptor-Mediated Apoptotic Mechanism in Caco-2 p53-Deficient Colon Adenocarcinoma Cells. *PLoS ONE* **2016**, *11*, e0146178. [CrossRef]
13. Chintharlapalli, S.; Papineni, S.; Ramaiah, S.K.; Safe, S. Betulinic Acid Inhibits Prostate Cancer Growth through Inhibition of Specificity Protein Transcription Factors. *Cancer Res.* **2007**, *67*, 2816–2823. [CrossRef]
14. Shanmugam, M.K.; Dai, X.; Kumar, A.P.; Tan, B.K.; Sethi, G.; Bishayee, A. Oleanolic acid and its synthetic derivatives for the prevention and therapy of cancer: Preclinical and clinical evidence. *Cancer Lett.* **2014**, *346*, 206–216. [CrossRef] [PubMed]
15. Bonel-Pérez, G.C.; Pérez-Jiménez, A.; Gris-Cárdenas, I.; Parra-Pérez, A.; Lupiáñez, J.; Reyes-Zurita, F.J.; Siles, E.; Csuk, R.; Peragón, J.; Rufino-Palomares, E.E. Antiproliferative and Pro-Apoptotic Effect of Uvaol in Human Hepatocarcinoma HepG2 Cells by Affecting G0/G1 Cell Cycle Arrest, ROS Production and AKT/PI3K Signaling Pathway. *Molecules* **2020**, *25*, 4254. [CrossRef]

16. Montilla, M.P.; Agil, A.; Navarro, M.C.; Jiménez, M.I.; García-Granados, A.; Parra, A.; Cabo, M.M. Antioxidant Activity of Maslinic Acid, a Triterpene Derivative Obtained from *Olea europaea*. *Planta Med.* **2003**, *69*, 472–474. [CrossRef]
17. Mokhtari, K.; Rufino-Palomares, E.E.; Pérez-Jiménez, A.; Reyes-Zurita, F.J.; Figuera, C.; García-Salguero, L.; Medina, P.P.; Peragón, J.; Lupiáñez, J.A. Maslinic Acid, a Triterpene from Olive, Affects the Antioxidant and Mitochondrial Status of B16F10 Melanoma Cells Grown under Stressful Conditions. *Evid. Based Complement. Altern. Med.* **2015**, *2015*, 272457. [CrossRef] [PubMed]
18. Mokhtari, K.; Pérez-Jiménez, A.; García-Salguero, L.; Lupiáñez, J.A.; Rufino-Palomares, E.E. Unveiling the Differential Antioxidant Activity of Maslinic Acid in Murine Melanoma Cells and in Rat Embryonic Healthy Cells Following Treatment with Hydrogen Peroxide. *Molecules* **2020**, *25*, 4020. [CrossRef] [PubMed]
19. Zhang, X.; Cao, J.; Zhong, L. Hydroxytyrosol inhibits pro-inflammatory cytokines, iNOS, and COX-2 expression in human monocytic cells. *Naunyn Schmiedeberg's Arch. Pharmacol.* **2009**, *379*, 581–586. [CrossRef] [PubMed]
20. Qian, Y.; Guan, T.; Tang, X.; Huang, L.; Huang, M.; Li, Y.; Sun, H. Maslinic acid, a natural triterpenoid compound from Olea europaea, protects cortical neurons against oxygen–glucose deprivation-induced injury. *Eur. J. Pharmacol.* **2011**, *670*, 148–153. [CrossRef] [PubMed]
21. Shaik, A.H.; Rasool, S.; Kareem, M.A.; Krushna, G.S.; Akhtar, P.M.; Devi, K.L. Maslinic Acid Protects Against Isoproterenol-Induced Cardiotoxicity in Albino Wistar Rats. *J. Med. Food* **2012**, *15*, 741–746. [CrossRef] [PubMed]
22. Pérez-Jiménez, A.; Rufino-Palomares, E.E.; Fernández-Gallego, N.; Ortuño-Costela, M.C.; Reyes-Zurita, F.J.; Peragón, J.; García-Salguero, L.; Mokhtari, K.; Medina, P.P.; Lupiáñez, J.A. Target molecules in 3T3-L1 adipocytes differentiation are regulated by maslinic acid, a natural triterpene from *Olea europaea*. *Phytomedicine* **2016**, *23*, 1301–1311. [CrossRef] [PubMed]
23. Parra, A.; Martin-Fonseca, S.; Rivas, F.; Reyes-Zurita, F.J.; Medina-O'Donnell, M.; Martinez, A.; García-Granados, A.; Lupiáñez, J.A.; Albericio, F. Semi-synthesis of acylated triterpenes from olive-oil industry wastes for the development of anticancer and anti-HIV agents. *Eur. J. Med. Chem.* **2014**, *74*, 278–301. [CrossRef]
24. Medina-O'Donnell, M.; Rivas, F.; Reyes-Zurita, F.J.; Cano-Muñoz, M.; Martinez, A.; Lupiáñez, J.A.; Parra, A. Oleanolic Acid Derivatives as Potential Inhibitors of HIV-1 Protease. *J. Nat. Prod.* **2019**, *82*, 2886–2896. [CrossRef] [PubMed]
25. Moneriz, C.; Mestres, J.; Bautista, J.M.; Diez, A.; Puyet, A. Multi-targeted activity of maslinic acid as an antimalarial natural compound. *FEBS J.* **2011**, *278*, 2951–2961. [CrossRef] [PubMed]
26. Fernández-Navarro, M.; Peragón, J.; Amores, V.; De La Higuera, M.; Lupiáñez, J.A. Maslinic acid added to the diet increases growth and protein-turnover rates in the white muscle of rainbow trout (*Oncorhynchus mykiss*). *Comp. Biochem. Physiol. Part C Toxicol. Pharmacol.* **2008**, *147*, 158–167. [CrossRef] [PubMed]
27. Rufino-Palomares, E.E.; Reyes-Zurita, F.J.; Fuentes-Almagro, C.; De La Higuera, M.; Lupiáñez, J.A.; Peragón, J. Proteomics in the liver of gilthead sea bream (*Sparus aurata*) to elucidate the cellular response induced by the intake of maslinic acid. *Proteomics* **2011**, *11*, 3312–3325. [CrossRef]
28. Rufino-Palomares, E.; Reyes-Zurita, F.; García-Salguero, L.; Peragón, J.; De La Higuera, M.; Lupiáñez, J.A. Maslinic acid, a natural triterpene, and ration size increased growth and protein turnover of white muscle in gilthead sea bream (*Sparus aurata*). *Aquac. Nutr.* **2012**, *18*, 568–580. [CrossRef]
29. Reyes-Zurita, F.J.; Rufino-Palomares, E.E.; Lupiáñez, J.A.; Cascante, M. Maslinic acid, a natural triterpene from *Olea europaea* L., induces apoptosis in HT29 human colon-cancer cells via the mitochondrial apoptotic pathway. *Cancer Lett.* **2009**, *273*, 44–54. [CrossRef] [PubMed]
30. Reyes-Zurita, F.J.; Pachón-Peña, G.; Lizárraga, D.; Rufino-Palomares, E.E.; Cascante, M.; Lupiáñez, J.A. The natural triterpene maslinic acid induces apoptosis in HT29 colon cancer cells by a JNK-p53-dependent mechanism. *BMC Cancer* **2011**, *11*, 154. [CrossRef] [PubMed]
31. Barroso, J.B.; García-Salguero, L.; Peragón, J.; De la Higuera, M.; Lupiáñez, J.A. The influence of dietary protein on the kinetics of NADPH production systems in various tissues of rainbow trout (*Oncorhynchus mykiss*). *Aquaculture* **1994**, *124*, 47–59. [CrossRef]
32. Lupiáñez, J.A.; Adroher, F.J.; Vargas, A.M.; Osuna, A. Differential behaviour of glucose 6-phosphate dehydrogenase in two morphological forms of Trypanosoma cruzi. *Int. J. Biochem.* **1987**, *19*, 1085–1089. [CrossRef]
33. Adroher, F.J.; Osuna, A.; Lupiáñez, J.A. Differential energetic metabolism during Trypanosoma cruzi differentiation. II. Hexokinase, phosphofructokinase and pyruvate kinase. *Mol. Cell. Biochem.* **1990**, *94*, 71–82. [CrossRef] [PubMed]
34. Sánchez-Muros, M.J.; García-Rejón, L.; Lupiáñez, J.A.; De la Higuera, M. Long-term nutritional effects on the primary liver and kidney metabolism in rainbow trout (*Oncorhynchus mykiss*). II. Adaptive response of glucose 6-phosphate dehydrogenase activity to high-carbohydrate/low-protein and high-fat/non-carbohydrate diets. *Aquac. Nutr.* **1996**, *2*, 193–200. [CrossRef]
35. Barroso, J.B.; Peragón, J.; García-Salguero, L.; De la Higuera, M.; Lupiáñez, J.A. Carbohydrate deprivation reduces NADPH-production in fish liver but not in adipose tissue. *Int. J. Biochem. Cell Biol.* **2001**, *33*, 785–796. [CrossRef]
36. Sheehan, D.; Meade, G.; Foley, V.M.; Dowd, C.A. Structure, function and evolution of glutathione transferases: Implications for classification of non-mammalian members of an ancient enzyme superfamily. *Biochem. J.* **2001**, *360*, 1–16. [CrossRef] [PubMed]
37. Corpas, F.J.; Barroso, J.B.; Sandalio, L.M.; Palma, J.M.; Lupiáñez, J.A.; del Río, L.A. Peroxisomal NADP-Dependent Isocitrate Dehydrogenase. Characterization and Activity Regulation during Natural Senescence. *Plant Physiol.* **1999**, *121*, 921–928. [CrossRef]
38. Parra, A.; Martin-Fonseca, S.; Rivas, F.; Reyes-Zurita, F.J.; Medina-O'Donnell, M.; Rufino-Palomares, E.E.; Martínez, A.; Garcia-Granados, A.; Lupiáñez, J.A.; Albericio, F. Solid-Phase Library Synthesis of Bi-Functional Derivatives of Oleanolic and Maslinic Acids and Their Cytotoxicity on Three Cancer Cell Lines. *ACS Comb. Sci.* **2014**, *16*, 428–447. [CrossRef]

39. Reyes-Zurita, F.J.; Medina-O'Donnell, M.; Ferrer-Martin, R.M.; Rufino-Palomares, E.E.; Martin-Fonseca, S.; Rivas, F.; Martínez, A.; García-Granados, A.; Pérez-Jiménez, A.; García-Salguero, L.; et al. The oleanolic acid derivative, 3-O-succinyl-28-O-benzyl oleanolate, induces apoptosis in B16–F10 melanoma cells via the mitochondrial apoptotic pathway. *RSC Adv.* **2016**, *6*, 93590–93601. [CrossRef]
40. Medina-O'Donnell, M.; Rivas, F.; Reyes-Zurita, F.J.; Martinez, A.; Martin-Fonseca, S.; Garcia-Granados, A.; Ferrer-Martín, R.M.; Lupiáñez, J.A.; Parra, A. Semi-synthesis and antiproliferative evaluation of PEGylated pentacyclic triterpenes. *Eur. J. Med. Chem.* **2016**, *118*, 64–78. [CrossRef]
41. Medina-O'Donnell, M.; Rivas, F.; Reyes-Zurita, F.J.; Martinez, A.; Galisteo-González, F.; Lupiáñez, J.A.; Parra, A. Synthesis and in vitro antiproliferative evaluation of PEGylated triterpene acids. *Fitoterapia* **2017**, *120*, 25–40. [CrossRef] [PubMed]
42. Medina-O'Donnell, M.; Rivas, F.; Reyes-Zurita, F.J.; Martinez, A.; Lupiáñez, J.A.; Parra, A. Diamine and PEGylated-diamine conjugates of triterpenic acids as potential anticancer agents. *Eur. J. Med. Chem.* **2018**, *148*, 325–336. [CrossRef] [PubMed]
43. Jannus, F.; Medina-O'Donnell, M.; Rivas, F.; Díaz-Ruiz, L.; Rufino-Palomares, E.E.; Lupiáñez, J.A.; Parra, A.; Reyes-Zurita, F.J. A Diamine-PEGylated Oleanolic Acid Derivative Induced Efficient Apoptosis through a Death Receptor and Mitochondrial Apoptotic Pathway in HepG2 Human Hepatoma Cells. *Biomolecules* **2020**, *10*, 1375. [CrossRef] [PubMed]
44. Jannus, F.; Medina-O'Donnell, M.; Neubrand, V.N.; Marín, M.; Sáez-Lara, M.J.; Sepúlveda, M.R.; Rufino-Palomares, E.E.; Martínez, A.; Lupiáñez, J.A.; Parra, A.; et al. Efficient In Vitro and In Vivo Anti-Inflammatory Activity of a Diamine-PEGylated Oleanolic Acid Derivative. *Int. J. Mol. Sci.* **2021**, *22*, 8158. [CrossRef]
45. Vega-Granados, K.; Medina-O'Donnell, M.; Rivas, F.; Reyes-Zurita, F.J.; Martinez, A.; Álvarez de Cienfuegos, L.; Lupiáñez, J.A.; Parra, A. Synthesis and Biological Activity of Triterpene–Coumarin Conjugates. *J. Nat. Prod.* **2021**, *84*, 1587–1597. [CrossRef] [PubMed]
46. Kang, W.S.; Jung, E.; Kim, J. Aucuba japonica Extract and Aucubin Prevent Desiccating Stress-Induced Corneal Epithelial Cell Injury and Improve Tear Secretion in a Mouse Model of Dry Eye Disease. *Molecules* **2018**, *23*, 2599. [CrossRef] [PubMed]
47. de Paz-Lugo, P.; Lupiáñez, J.A.; Meléndez-Hevia, E. High glycine concentration increases collagen synthesis by articular chondrocytes in vitro—Acute glycine deficiency could be an important cause of osteoarthritis. *Amino Acids* **2018**, *50*, 1357–1365. [CrossRef]
48. de Paz-Lugo, P.; Meléndez-Hevia, E. Branch-point stoichiometry can generate weak links in metabolism: The case of glycine biosynthesis. *J. Biosci.* **2008**, *33*, 771–780. [CrossRef]
49. Hong, G.U.; Lee, J.Y.; Kang, H.; Kim, T.Y.; Park, J.Y.; Hong, E.Y.; Shin, Y.H.; Jung, S.H.; Chang, H.-B.; Kim, Y.H.; et al. Inhibition of Osteoarthritis-Related Molecules by Isomucronulatol 7-O-β-d-glucoside and Ecliptasaponin A in IL-1β-Stimulated Chondrosarcoma Cell Model. *Molecules* **2018**, *23*, 2807. [CrossRef] [PubMed]
50. Hong, E.Y.; Kim, T.Y.; Hong, G.U.; Kang, H.; Lee, J.Y.; Park, J.Y.; Kim, S.C.; Kim, Y.H.; Chung, M.H.; Kwon, Y.I.; et al. Inhibitory Effects of Roseoside and Icariside E4 Isolated from a Natural Product Mixture (No-ap) on the Expression of Angiotensin II Receptor 1 and Oxidative Stress in Angiotensin II-Stimulated H9C2 Cells. *Molecules* **2019**, *24*, 414. [CrossRef]
51. Loesche, A.; Kahnt, M.; Serbian, I.; Brandt, W.; Csuk, R. Triterpene-Based Carboxamides Act as Good Inhibitors of Butyrylcholinesterase. *Molecules* **2019**, *24*, 948. [CrossRef]
52. Panicker, N.G.; Balhamar, S.O.M.S.; Akhlaq, S.; Qureshi, M.M.; Rizvi, T.S.; Al-Harrasi, A.; Hussain, J.; Mustafa, F. Identification and Characterization of the Caspase-Mediated Apoptotic Activity of *Teucrium mascatense* and an Isolated Compound in Human Cancer Cells. *Molecules* **2019**, *24*, 977. [CrossRef] [PubMed]
53. Juan, M.E.; Lozano-Mena, G.; Sánchez-González, M.; Planas, J.M. Reduction of Preneoplastic Lesions Induced by 1,2-Dimethylhydrazine in Rat Colon by Maslinic Acid, a Pentacyclic Triterpene from *Olea europaea* L. *Molecules* **2019**, *24*, 1266. [CrossRef] [PubMed]
54. Lv, Y.; Hou, X.; Zhang, Q.; Li, R.; Xu, L.; Chen, Y.; Tian, Y.; Sun, R.; Zhang, Z.; Xu, F. Untargeted Metabolomics Study of the In Vitro Anti-Hepatoma Effect of Saikosaponin d in Combination with NRP-1 Knockdown. *Molecules* **2019**, *24*, 1423. [CrossRef] [PubMed]

molecules

MDPI

Article

Aucuba japonica Extract and Aucubin Prevent Desiccating Stress-Induced Corneal Epithelial Cell Injury and Improve Tear Secretion in a Mouse Model of Dry Eye Disease

Wan Seok Kang [1,†], Eunsoo Jung [2,†] and Junghyun Kim [1,*]

1 College Department of Oral Pathology, School of Dentistry, Chonbuk National University, Jeonju 54896, Korea; wskangjbnu@gmail.com

2 Laboratory of Toxicology, Research Institute for Veterinary Science and College of Veterinary Medicine, Seoul National University, Seoul 08826, Korea; ozz79@snu.ac.kr

* Correspondence: dvmhyun@jbnu.ac.kr; Tel.: +82-63-270-4032; Fax: +82-63-270-4025

† These authors contributed equally to this paper.

Received: 21 September 2018; Accepted: 8 October 2018; Published: 11 October 2018

check for updates

Abstract: Dry eye disease is affected by a broad range of causes such as age, lifestyle, environment, medication and autoimmune diseases. These causes induce tear instability that activates immune cells and promotes expression of inflammatory molecules. In this study, we investigated the therapeutic effects of an ethanolic extract of *Aucuba japonica* (AJE) and its bioactive compound, aucubin, on dry eye disease. The human corneal cells were exposed to desiccation stress induced by exposing cells to air, so that viability was decreased. On the other hand, pre-treatment of AJE and aucubin restored cell survival rate depending on the dose under the dry condition. This result was confirmed again by terminal deoxynucleotidyl transferase dUTP nick end labeling (TUNEL) staining. The mRNA expression of inflammatory molecules was reduced by the pretreatment of AJE and aucubin under the dry state. The therapeutic effects of AJE and aucubin were examined in the animal model for dry eye induced by unilateral excision of the exorbital lacrimal gland. Declined tear volumes and corneal irregularity in the dry eye group were fully recovered by the administration of AJE and aucubin. The apoptotic cells on the cornea were also decreased by AJE and aucubin. Therefore, this study suggests that administration of AJE can be a novel therapeutic for dry eye disease and that the pharmacological activities of AJE may be in part due to its bioactive compound, aucubin.

Keywords: *Aucuba japonica*; aucubin; cornea; dry eye; tear

---

1. Introduction

Dry eye disease (DED) is a chronic pathogenic condition, which is characterized by a lack of tear production or an increase in tear evaporation, usually caused by ocular surface inflammation [1]. Most older people, but not exclusively them, are suffering DED; many of causes can affect the onset of it including intrinsic factors such as systemic medication, autoimmune disorders, or extrinsic factors such as low humidity of the environment, prolonged video display viewing and laser-assisted in situ keratomileusis (LASIK) surgery [2,3]. When the patients experience dysfunction of tear secretion, they already have a decreased function of secretory cells or glands resulting in a formation of the unstable tear film. Tear instability exerts the activation of stress-related signaling pathways of epithelial cells and resident inflammatory cells in the ocular surface, which induces the production of pro-inflammatory cytokines [4]. It may exacerbate the deficiency of tear production and worsen dry eye symptoms. Therefore, inhibition of inflammation and apoptosis can be a promising therapeutic strategy for DED.

Aucuba japonica, a perennial native plant of Korea and Japan, has been used as food and as a medicinal plant to treat several diseases including edema, abscess and gastrointestinal disorders [5]. The leaves have various chemical compounds such as aucubin and eucommiol-*O*-β-D-glucopyranoside [6]; in particular, aucubin is abundant, a small molecule that is known for anti-inflammatory and anti-oxidant effects [7,8], liver protection [9], gastroprotection [10], neuro-protection [8] and promoting angiogenesis [11]. Although *Aucuba japonica* and its bioactive ingredient, aucubin, have shown various pharmacological activities in vitro and in vivo, its pharmacological activities on DED still have not been evaluated. The purpose of this study was to examine the effects of an extract of *Aucuba japonica* and aucubin on desiccating stress-induced corneal epithelial cell injury in vitro and tear film instability in a mouse model induced by unilateral excision of the exorbital lacrimal gland and identify the possible mechanism of action.

2. Results

2.1. HPLC Analysis of AJE

As shown in Figure 1, aucubin was found in AJE. The content of aucubin in AJE was 284.30 ± 1.54 mg/g (Table 1). According to the content of aucubin in AJE, we determined the doses of aucubin for in vitro and in vivo experiments.

Table 1. Content of aucubin in the *Aucuba japonica* extract.

Compound	Content (mean ± SE, n = 3), mg/g (%)
Aucubin	284.30 ± 0.89 (28.4)

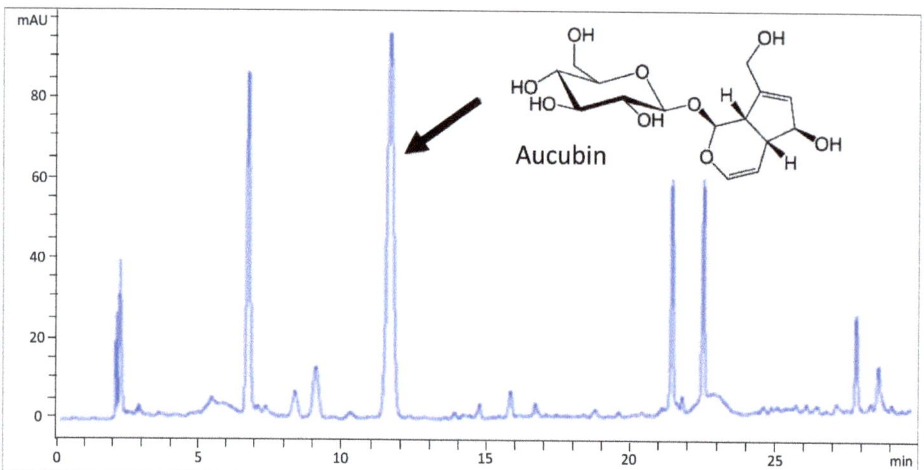

Figure 1. HPLC chromatographs of an extract of *Aucuba japonica*.

2.2. AJE and Aucubin Improve Cell Viability against Desiccation Injury

The cytotoxicity of AJE and aucubin to human corneal epithelial cells was examined by MTS assay. We tested 1 to 100 μg/mL range of AJE and 0.1 to 30 μg/mL range of aucubin. There was no cytotoxicity (Figure 2A). Desiccation was used as a stress corresponding to dry eye disease. The cells were exposed to air on the clean bench for 10 to 30 min, and 25 min was used as a moderate damage condition (Figure 2B). To investigate the preventive effects of AJE and aucubin against desiccation, the cells were pre-treated various concentrations of AJE and aucubin for 24 h. The cell viability was significantly recovered, more than 70% in both of 25 and 50 µg/mL of AJE- treated groups. Aucubin also restored the cell viability in a dose-dependent manner.

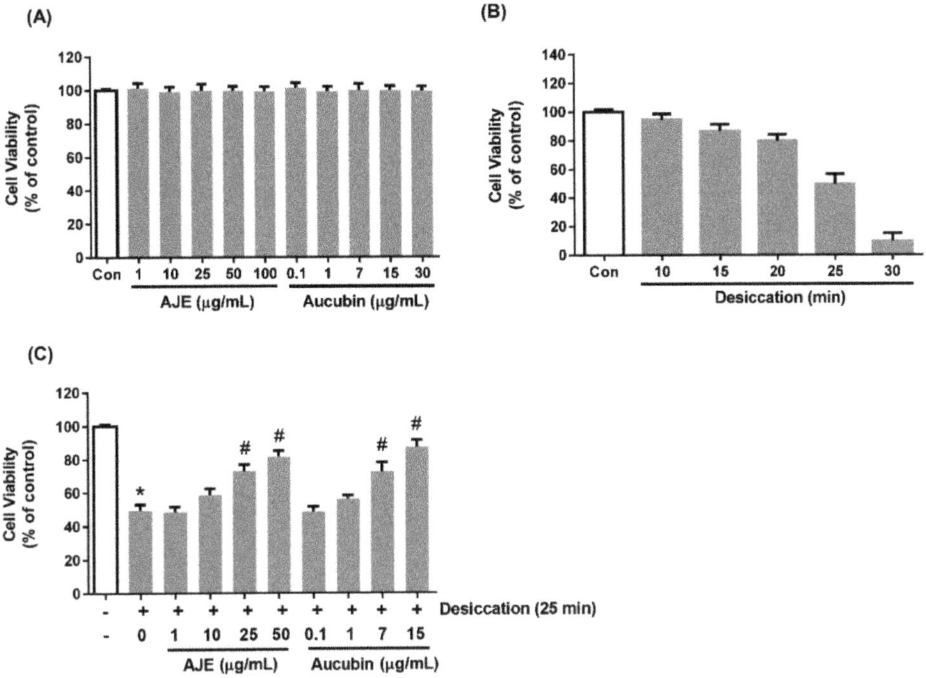

Figure 2. Effects of *Aucuba japonica* (AJE) and aucubin on the viability of human corneal epithelial cell. (**A**) The cell viability in the dose-dependent treatments of AJE and aucubin for 24 h. (**B**) The cell viability under the time-dependent desiccation. (**C**) The cells were pre-incubated with various concentrations of AJE and aucubin for 24 h and then exposed to air to induce desiccation stress. All of the viability is shown as % of control. The values in bar graph represent the mean ± SE. * $p < 0.05$ vs. control cell, # $p < 0.05$ vs. vehicle treated cell.

2.3. *AJE and Aucubin Inhibit Apoptosis against Desiccation Stress*

To confirm the cytoprotective effects of AJE and aucubin more directly, we examined apoptosis assay. The apoptotic cells were detected by TUNEL staining. The dying cells were increased by desiccation process and the ratio of apoptotic cells was significantly decreased in both of 50 µg/mL AJE-treated group and 15 µg/mL aucubin-treated group (Figure 3A,B). These results gave us a possibility of use of AJE and aucubin as anti-dry eye disease reagent.

2.4. *AJE and Aucubin Decrease mRNA of Inflammatory Cytokine Expression*

Dry eye disease is further encouraged by inflammation [4]. Therefore, we evaluated the expression of inflammatory genes to determine whether AJE and aucubin showed anti-inflammatory effects under dry conditions. The mRNA expression of IL-1β, IL-8 and TNF-α were significantly increased in the dry condition, and those were lowered by AJE and aucubin treatments (Figure 4A–C).

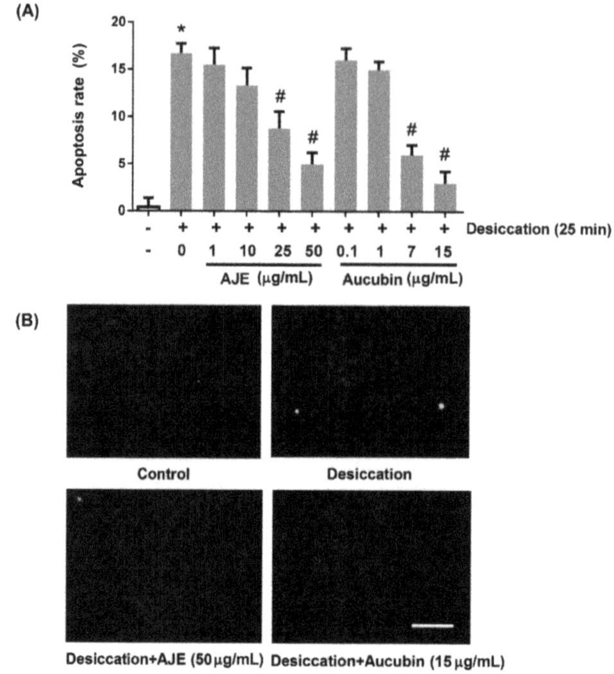

Figure 3. Effects of AJE and aucubin on apoptosis rate of human corneal epithelial cell under desiccation. The cell was pre-incubated with various concentrations of AJE and aucubin for 24 h and then exposed to air to induce desiccation stress. After that, the cells were fixed and performed TUNEL staining. (**A**) Apoptosis rate was calculated by comparing the ratio of TUNEL-positive cells to total cell number. (**B**) The images were taken under the fluorescence microscope after TUNEL staining. Scale bars = 100 μm. The values in bar graph represent the mean ± SE. * $p < 0.05$ vs. control cell, # $p < 0.05$ vs. vehicle treated cell.

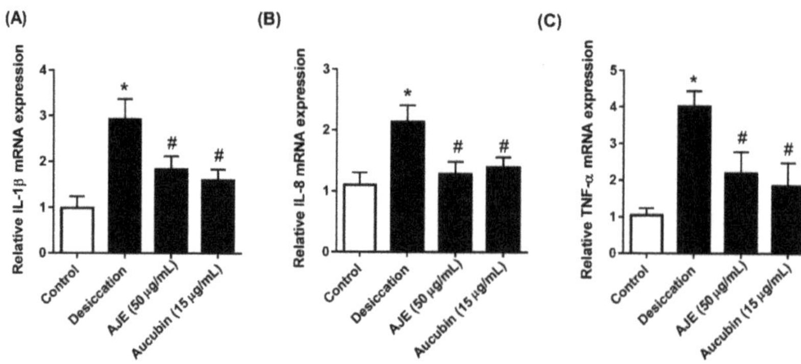

Figure 4. Effects of AJE and aucubin on mRNA level of inflammatory cytokines in human corneal epithelial cell. The cell was pre-incubated with 50 μg/mL of AJE and 15 μg/mL of aucubin for 24 h and then exposed to air to induce desiccation stress. Total RNA was extracted and the mRNA of (**A**) IL-1β, (**B**) IL-8 and (**C**) TNF-α were analyzed by real-time PCR. The data in bar graph are relative values compare to control and represent the mean ± SE. * $p < 0.05$ vs. control cell, # $p < 0.05$ vs. vehicle treated cell.

2.5. AJE and Aucubin Improve Tear Production and Corneal Irregularity in the DED Rats

In order to test AJE and aucubin on in vivo dry eye rat model, we induced experimental DED by surgical excision of the unilateral exorbital lacrimal gland. The AJE and aucubin were orally administered for 7 days. Tear production was determined using a phenol red-impregnated cotton thread and the wet lengths of thread indicated tear volume were compared. The experimental dry eye model showed significantly lower tear volume compared with the normal group, but those were recovered by AJE and aucubin treatments. In particular, the higher dose of AJE greatly recovered tear volume to almost normal. (Figure 5A). The corneal irregularity was assessed by taking pictures of the rat eye reflecting a white ring of light. The dry eye group showed an irregular circle shape of light, but AJE and aucubin treatments improved this change (Figure 5B). The scores of corneal irregularities showed that the dry eye group had a high score, but that AJE and aucubin treatments decreased it significantly (Figure 5C).

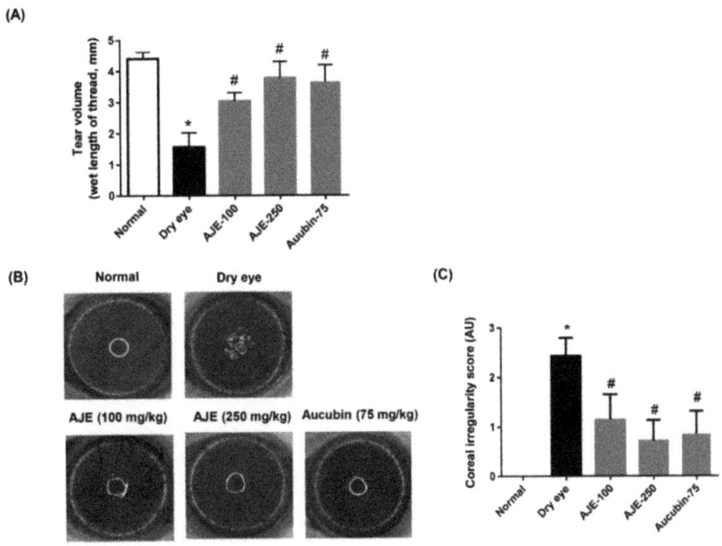

Figure 5. Effects of AJE and aucubin on tear secretion and corneal irregularity in dry eye rats. (**A**) Tear volume was measured using the phenol red thread test. Tear volume was expressed in millimeters of thread that became wet by the tear and turned red in color. (**B**) Reflected images of a white ring from the fiberoptic ring illuminator of a stereomicroscope. (**C**) Corneal irregularity was graded according to the number of distorted quarters in the reflected white ring as follows: 0, no distortion; 1, distortion in one quarter; 2, distortion in two quarters; 3, distortion in three quarters; 4, distortion in all four quarters; 5, severe distortion, in which no ring could be recognized. The values in bar graph represent the mean ± SE. n = 7. * $p < 0.05$ vs. normal rat, # $p < 0.05$ vs. vehicle treated dry eye rat.

2.6. AJE and Aucubin Reduce Apoptotic Cells on the Cornea of Dry Eye Rats

As shown in the above data, in order to confirm whether the cytoprotective effect of AJE and aucubin actually work on an in vivo dry eye model, we performed TUNEL staining to detect apoptotic cells on cornea of rats. The apoptotic cells on the cornea were significantly increased in dry eye rats than those of normal rats, but those were reduced by AJE and aucubin treatments (Figure 6A,B).

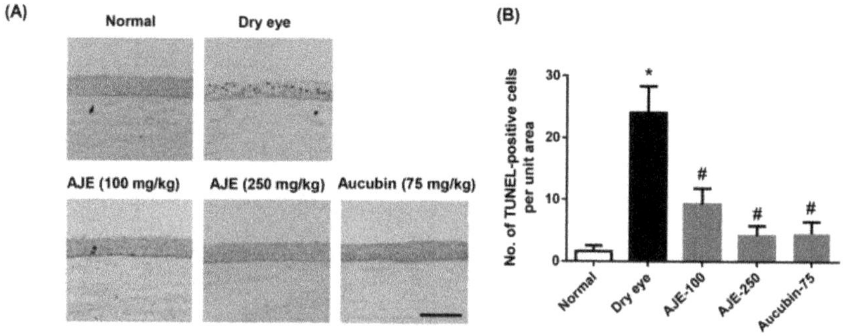

Figure 6. Effects of AJE and aucubin on the number of apoptotic cells in ocular surface of dry eye rats. (**A**) Apoptotic cells were stained by TUNEL assay. Scale bars = 100 μm. (**B**) TUNEL-positive cells were counted in the unit area. The values in bar graph represent the mean ± SE. n = 7. * $p < 0.05$ vs. normal rat, # $p < 0.05$ vs. vehicle treated dry eye rat.

3. Discussion

DED is a disorder of the tear film due to tear deficiency or excessive evaporation [12,13]. Numerous risk factors for DED have been reported, such as aging, gender, contact lens use, air-conditioning and visual display terminal work [14–17]. Deficiency of tears is caused by various pathophysiological mechanisms, including cell volume reduction, dysfunction to DNA repair systems, increased reactive oxygen species generation and increased apoptotic cell death in the ocular surface cells [18]. The most important core mechanism for DED is associated with hyperosmolarity of the tear film and ocular surface inflammation [19].

Current treatment strategies for dry eye disease are using artificial tears, lubricants and various ophthalmic drops, such as anti-inflammatory drugs, immunosuppressive agents and steroids. However, these treatments have focused on alleviating clinical symptoms rather than resolving the causes of DED. Therefore, in order to achieve a fundamental therapy of dry eye disease, we investigated the protective effect of a natural product, an extract of *Aucuba japonica*, that has traditionally been used for the treatment of burns, incised wounds and swelling. One of the major contents of the plant is aucubin, which has demonstrated multiple protective effects including anti-inflammation, anti-oxidation and tissue protection. We also examined whether aucubin is a major bioactive ingredient in the AJE.

Desiccation injury on cells used here was widely performed to discover therapeutic molecules or diseases progression mechanism [20–22]. AJE and aucubin treatments on corneal cell increased viability against desiccation injury in our data. It is the first preventive report using AJE and aucubin for the dry condition, even though a previous study showed similar data, that aucubin has a protective effect on H_2O_2 induced apoptosis [8]. This result was confirmed by TUNEL staining. As reported previously, the anti-inflammatory effect of aucubin was also detected by our AJE and aucubin treatments on dry injured cells. Aucubin prevents proinflammatory cytokine-induced inflammatory response by inhibition of NF-κB signaling [23,24]. Aucubin is also reported to regulate Bcl-2 family protein expression and inhibit apoptosis [8,25]. Ho et al. reported that aucubin protected against UVB-induced skin fibroblast activation by inhibition of the production of matrix metalloproteinase-1 [26]. Since inflammation and corneal epithelial cell injury are involved in the progression of DED, aucubin was predicted to have protective effects on DED. These anti-apoptotic and anti-inflammatory effects of AJE and aucubin may work synergistically to prevent desiccation injury, even though we did not confirm detailed mechanisms.

In the animal experiments, AJE and aucubin were orally administrated in the DED rats for 7 days. The tear volume and corneal irregularity were recovered almost to normal by AJE and aucubin treatment in our data. AJE and aucubin also decreased apoptotic cell injury of the cornea of dry eye rats. Our results suggest that the cytoprotective effects of AJE and its bioactive compound, aucubin, are mediated by its anti-apoptotic and anti-inflammatory activities. Inflammation acts as a major factor of DED. We did not test that AJE and aucubin directly induce tear secretion or expression of tear components, but we believe that it would be a good fundamental therapeutic, because AJE and aucubin clearly inhibit the loss of cells consisting of the cornea. Therefore, further study should be performed to confirm the effectiveness, as well as detailed therapeutic mechanisms and safety.

In conclusion, AJE and aucubin can inhibit corneal cell apoptosis induced by dry conditions. Progression of the dry eye disease is clearly suppressed by AJE and aucubin treatments, and dysfunction of tear secretion is also protected. Therefore, the current work demonstrates that AJE and aucubin may pave the pathway to treat dry eye disease.

4. Experimental Section

4.1. Preparation of *Aucuba japonica* Extract

The leaves and stems of cultivated *Aucuba japonica* were purchased from Namhae, Kyungsangnamdo, South Korea. The botanical identification was made by botanist Prof. Joo-Hwan Kim (Department of life science, Gachon University, Kyounggi-do, South Korea). The voucher specimen (No. JBNU-AJE) is deposited in the herbarium of the Korea Institute of Oriental Medicine (Daejoen, South Korea). To prepare the standardized *A. japonica* extract (AJE), 1 kg of the aerial parts of *A. japonica* was weighed accurately, and 30% ethanol (10 L) was added to the herb and extracted at 70 °C for 3 h. The extract solution was filtered and concentrated to get a 210 g extract. AJE was standardized by high-performance liquid chromatography (HPLC) using aucubin (Sigma, St. Louis, MO, USA) as a reference compound. According to the content of aucubin in AJE, we determined the doses of aucubin for in vitro and in vivo experiments.

4.2. Cell Culture and Cell Viability Assay

The human corneal epithelial cells (PCS-700-010, American Type Culture Collection, Manassas, VA, USA) were maintained in corneal epithelial cell basal medium containing growth supplements according to the manufacturer's instructions (American Type Culture Collection, Manassas, VA, USA). Cell viability was examined using an MTS assay kit (Promega Corporation, Madison, WI, USA). Cells (1×10^4 cells/well) were plated in 96-well plates. Cells at a density of 2×10^5/well were treated with various concentrations of AJE (1–100 μg/mL) or aucubin (0.1–30 μg/mL). AJE and aucubin were dissolved in the culture medium. The content of aucubin in AJE was 284.30 ± 1.54 mg/g. According to the content of aucubin in AJE, we determined the doses of aucubin. Cell viability was measured 24 h after incubation. The results of the MTS assay were obtained by measuring absorbance using a microplate reader (Tecan Group Ltd., Männedorf, Switzerland) at 490 nm.

4.3. Cell Viability and Apoptosis under Desiccating Stress

Desiccating stress on the cell was induced after pre-incubation with AJE or aucubin for 24 h. To induce desiccating stress, the culture media was discarded and cells were exposed to different durations of desiccation (0, 10, 15, 20, 25, and 30 min). The viability of desiccated cell was measured by MTS assay as described in above. In order to detect apoptotic cells after desiccation, the cells were seed on the autoclaved coverslip on the plate. After pre-incubation with AJE for 24 h, the cells were exposed in a dry condition for 25 min and apoptotic cells were detected by TUNEL staining.

4.4. Real-Time PCR

Total RNA from the corneal epithelial cells was extracted using Trizol reagent (Invitrogen, Carlsbad, CA, USA), and cDNA was synthesized using M-MLV reverse transcriptase (Bioneer, Daejeon, Korea). PCRs were performed using an iQ5 Continuous Fluorescence Detector System (Bio-Rad, CA, USA), with 2xSYBR®Green PCR Master

Mix (SYBR Premix Ex Taq TM, TaKaRa, Tokyo, Japan). All procedures were conducted according to the manufacturer's instruction. The sequences of the primers were as follows: IL-1b sense 5′-AGCTACGAATCTCCGACCAC-3′; IL-1b antisense 5′-CGTTATCCCATGTGTCGAAGAA-3′; IL-8 sense 5′-ATGCTTTTGATCTGCACAGCTGCAC-3′; IL-8 antisense 5′-TGGTCCAGCAGGAATAACCCTCAG-3′; TNF-α sense 5′-CAGCCTCTTCTCCTTCCTGA-3′; TNF-α antisense 5′-GGAAGACCCCTCCCAGATAGA-3′; GAPDH sense 5′-ATGTTCGTCATGGGTGTGAA-3′, and GAPDH antisense 5′-GGTGCTAAGCAGTTGGTGGT-3′. The results were normalized to GAPDH.

4.5. Animal Experiments

Seven-week-old male SD rats (Orient Bio, Seoul, Korea) were deeply anesthetized with isoflurane (JW Pharmaceutical, Seoul, Korea). To induce experimental DED, the left exorbital lacrimal gland was surgically excised. At three days after surgery, the animals were divided randomly into four groups (n = 7/group): (1) vehicle-treated DED rat (DED); (2) 100 mg/kg AJE-treated DED rats (AJE-100); (3) 250 mg/kg AJE-treated DED rats (AJE-250); (4) 75 mg/kg aucubin-treated DED rats (Aucubin-75). AJE and aucubin were dissolved in distilled water and orally administered for 7 days. Seven rats in the normal group (NOR) received a sham operation. All procedures performed on the animals were approved by our Institutional Animal Care and Use Committee (IACUC approval No. 17-059).

4.6. Tear Volume Measurement

After a 7-day treatment of AJE, tear volume was measured using a phenol red thread (Zone Quick, FCI Ophthalmics, Pembroke, MA, USA). The cotton thread was placed in the lateral canthus for 30 s. The length of the color-changed thread was measured. Measurements of tear volume were obtained in both eyes.

4.7. Corneal Surface Irregularity

To evaluate desiccating stress-induced ocular surface change, the corneal surface irregularity was evaluated as previously described [18]. The irregularity of reflected light from the fiberoptic ring illuminator (SZ51; Olympus, Tokyo, Japan) on the corneal surface were scored as follows: 0, no distortion; 1, distortion in 1/4 quarter of the reflected ring shape; 2, distortion in 2/4 quarters; 3, distortion in 3/4 quarters; 4, distortion in 4/4 quarters; 5, severe distortion and no ring shape could be recognized.

4.8. TUNEL Staining

The apoptotic cells on the tissue were determined by terminal deoxyribonucleotidyl transferase (TdT)-mediated dUTP nick end labelling (TUNEL) staining according to the manufacturer's instructions.

4.9. Statistical Analysis

Data in all of the tables and figures are presented as the mean ± standard error of the mean (SE). Significant differences were assessed using one-way analysis of variance (ANOVA) followed by Tukey's multiple comparison test. Differences were considered statistically significant at $p < 0.05$.

Author Contributions: W.S.K. performed experiments and wrote the manuscript. E.J. performed experiments. J.K. designed and supervised the study. All authors read and approved the final manuscript.

Funding: This research was supported by the R&D Program for Forest Science Technology (Project No. 2017040A00-1719-BA01) provided by Korea Forest Service (Korea Forestry Promotion Institute) and the Korea Institute of Planning and Evaluation for Technology in Food, Agriculture, Forestry and Fisheries (IPET) through Agri-Bio industry Technology Development Program, funded by Ministry of Agriculture, Food and Rural Affairs (grant No: 116081-03-3-CG000). This research was also supported by research funds for newly appointed professors of Chonbuk National University in 2017.

Conflicts of Interest: The authors declare no conflict of interest.

References

1. Drew, V.J.; Tseng, C.-L.; Seghatchian, J.; Burnouf, T. Reflections on dry eye syndrome treatment: Therapeutic role of blood products. *Front. Med.* **2018**, *5*, 33. [CrossRef] [PubMed]
2. Pflugfelder, S.C.; de Paiva, C.S. The pathophysiology of dry eye disease: What we know and future directions for research. *Ophthalmology* **2017**, *124*, S4–S13. [CrossRef] [PubMed]
3. Yao, W.; Davidson, R.S.; Durairaj, V.D.; Gelston, C.D. Dry eye syndrome: An update in office management. *Am. J. Med.* **2011**, *124*, 1018. [CrossRef] [PubMed]
4. Pflugfelder, S.C.; Stern, M.E. Mucosal environmental sensors in the pathogenesis of dry eye. *Expert. Rev. Clin. Immunol.* **2014**, *10*, 1137–1140. [CrossRef] [PubMed]
5. Kimura, T.; But, P.P.H.; Guo, J.X.; Sung, C.K. International collation of traditional and folk medicine. *World Sci.* **1997**, *2*, 238. [CrossRef]
6. Bernini, R.; Iavarone, C.; Trogolo, C. 1-*O*-β-D-glucopyranosyleucommiol, an iridoid glucoside from *Aucuba japonica*. *Phytochemistry* **1984**, *23*, 1431–1433. [CrossRef]
7. Park, K.S.; Chang, I.-M. Anti-inflammatory activity of aucubin by inhibition of tumor necrosis factor-α production in RAW 264.7 cells. *Planta Med.* **2004**, *70*, 778–779. [CrossRef] [PubMed]
8. Xue, H.Y.; Gao, G.Z.; Lin, Q.Y.; Jin, L.J.; Xu, Y.P. Protective effects of aucubin on H_2O_2-induced apoptosis in PC12 cells. *Phytother. Res.* **2012**, *26*, 369–374. [CrossRef] [PubMed]
9. Chang, I.-M. Liver-protective activities of aucubin derived from traditional oriental medicine. *Res. Commun. Mol. Pathol. Pharmacol.* **1998**, *102*, 189–204. [PubMed]
10. Yang, Y.; Yin, B.; Lv, L.; Wang, Z.; He, J.; Chen, Z.; Wen, X.; Zhang, Y.; Sun, W.; Li, Y. Gastroprotective effect of aucubin against ethanol-induced gastric mucosal injury in mice. *Life Sci.* **2017**, *189*, 44–51. [CrossRef] [PubMed]
11. Chen, L.; Yang, Y.; Zhang, L.; Li, C.; Coffie, J.W.; Geng, X.; Qiu, L.; You, X.; Fang, Z.; Song, M. Aucubin promotes angiogenesis via estrogen receptor beta in a mouse model of hindlimb ischemia. *J. Steroid Biochem. Mol. Biol.* **2017**, *172*, 149–159. [CrossRef] [PubMed]
12. Lemp, M.A. Report of the national eye institute/industry workshop on clinical trials in dry eyes. *CLAO J.* **1995**, *21*, 221–232. [PubMed]
13. Tsubota, K.; Yokoi, N.; Shimazaki, J.; Watanabe, H.; Dogru, M.; Yamada, M.; Kinoshita, S.; Kim, H.M.; Tchah, H.W.; Hyon, J.Y.; et al. New perspectives on dry eye definition and diagnosis: A consensus report by the Asia dry eye society. *Ocul. Surf.* **2017**, *15*, 65–76. [CrossRef] [PubMed]
14. Uchino, M.; Nishiwaki, Y.; Michikawa, T.; Shirakawa, K.; Kuwahara, E.; Yamada, M.; Dogru, M.; Schaumberg, D.A.; Kawakita, T.; Takebayashi, T.; et al. Prevalence and risk factors of dry eye disease in Japan: Koumi study. *Ophthalmology* **2011**, *118*, 2361–2367. [CrossRef] [PubMed]
15. Uchino, M.; Yokoi, N.; Uchino, Y.; Dogru, M.; Kawashima, M.; Komuro, A.; Sonomura, Y.; Kato, H.; Kinoshita, S.; Schaumberg, D.A.; et al. Prevalence of dry eye disease and its risk factors in visual display terminal users: The Osaka study. *Am. J. Ophthalmol.* **2013**, *156*, 759–766. [CrossRef] [PubMed]
16. Schaumberg, D.A.; Sullivan, D.A.; Buring, J.E.; Dana, M.R. Prevalence of dry eye syndrome among US women. *Am. J. Ophthalmol.* **2003**, *136*, 318–326. [CrossRef]
17. Gonzalez-Meijome, J.M.; Parafita, M.A.; Yebra-Pimentel, E.; Almeida, J.B. Symptoms in a population of contact lens and noncontact lens wearers under different environmental conditions. *Optom. Vis. Sci.* **2007**, *84*, 296–302. [CrossRef] [PubMed]
18. Barabino, S.; Labetoulle, M.; Rolando, M.; Messmer, E.M. Understanding symptoms and quality of life in patients with dry eye syndrome. *Ocul. Surf.* **2016**, *14*, 365–376. [CrossRef] [PubMed]
19. Gaffney, E.A.; Tiffany, J.M.; Yokoi, N.; Bron, A.J. A mass and solute balance model for tear volume and osmolarity in the normal and the dry eye. *Prog. Retin. Eye Res.* **2010**, *29*, 59–78. [CrossRef] [PubMed]
20. Matsuo, T. Trehalose protects corneal epithelial cells from death by drying. *Columbia J. Ophthalmol.* **2001**, *85*, 610–612. [CrossRef]
21. Higuchi, A.; Kawakita, T.; Tsubota, K. IL-6 induction in desiccated corneal epithelium in vitro and in vivo. *Mol. Vis.* **2011**, *17*, 2400. [PubMed]

22. Tost, F.; Keiss, R.; Großjohann, R.; Jürgens, C.; Giebel, J. Effect of different artificial tears against desiccation in cultured human epithelial cells. *Med. Sci. Monit.* **2012**, *18*, BR188–BR192. [CrossRef] [PubMed]
23. Wang, S.N.; Xie, G.P.; Qin, C.H.; Chen, Y.R.; Zhang, K.R.; Li, X.; Wu, Q.; Dong, W.Q.; Yang, J.; Yu, B. Aucubin prevents interleukin-1 beta induced inflammation and cartilage matrix degradation via inhibition of NF-κB signaling pathway in rat articular chondrocytes. *Int. Immunopharmacol.* **2015**, *24*, 408–415. [CrossRef] [PubMed]
24. Park, K.S. Aucubin, a naturally occurring iridoid glycoside inhibits TNF-alpha-induced inflammatory responses through suppression of NF-κB activation in 3T3-L1 adipocytes. *Cytokine* **2013**, *62*, 407–412. [CrossRef] [PubMed]
25. Xue, H.Y.; Niu, D.Y.; Gao, G.Z.; Lin, Q.Y.; Jin, L.J.; Xu, Y.P. Aucubin modulates Bcl-2 family proteins expression and inhibits caspases cascade in H_2O_2-induced PC12 cells. *Mol. Biol. Rep.* **2011**, *38*, 3561–3567. [CrossRef] [PubMed]
26. Ho, J.N.; Lee, Y.H.; Lee, Y.D.; Jun, W.J.; Kim, H.K.; Hong, B.S.; Shin, D.H.; Cho, H.Y. Inhibitory effect of Aucubin isolated from Eucommia ulmoides against UVB-induced matrix metalloproteinase-1 production in human skin fibroblasts. *Biosci. Biotechnol. Biochem.* **2005**, *69*, 2227–2231. [CrossRef] [PubMed]

Sample Availability: Samples of the compounds at amounts less than 1 mg are available from the authors.

molecules

MDPI

Article

Inhibition of Osteoarthritis-Related Molecules by Isomucronulatol 7-*O*-β-D-glucoside and Ecliptasaponin A in IL-1β-Stimulated Chondrosarcoma Cell Model

Gwan Ui Hong [1,†], Jung-Yun Lee [2,†], Hanna Kang [2], Tae Yang Kim [2], Jae Yeo Park [1], Eun Young Hong [1], Youn Ho Shin [1], Sung Hoon Jung [1], Hung-Bae Chang [3], Young Ho Kim [4], Young-In Kwon [2,*] and Jai Youl Ro [1,5,*]

[1] Life & Science Research Center, Hyunsung Vital Co. Ltd., Seoul 07255, Korea; gwanuihong@gmail.com (G.U.H.); legolas018@naver.com (J.Y.P.); h_medical@naver.com (E.Y.H.); dush84@hanmail.net (Y.H.S.); jsh62548@naver.com (S.H.J.)

[2] Department of Food and Nutrition, Hannam University, Daejeon 34054, Korea; seembeeks@hanmail.net (J.-Y.L.); hanna9506@hanmail.net (H.K); xodid5606@naver.com (T.Y.K.)

[3] Department of Bio Quality Control, Korea Bio Polytechnic, Chungnam 32943, Korea; hbchang@kopo.ac.kr

[4] Department of Pharmacy, Choongnam National University, Daejeon 34134, Korea; yhk@cnu.ac.kr

[5] Department of Pharmacy, Sungkyunkwan University, Suwon 03063, Korea

* Correspondence: youngk@hnu.kr (Y.-I.K.); jyro426@skku.edu (J.Y.R.); Tel.: +82-42-629-8790 (Y.-I.K.); +82-2-2628-5105 (J.Y.R.); Fax: +82-42-629-8789 (Y.-I.K.); +82-2-2628-0572 (J.Y.R.)

† These authors contributed equally to this work.

Academic Editors: Eva E. Rufino-Palomares and José Antonio Lupiáñez

Received: 2 September 2018; Accepted: 26 October 2018; Published: 29 October 2018

check for updates

Abstract: Osteoarthritis (OA) is the common form of arthritis and is characterized by disability and cartilage degradation. Although natural product extracts have been reported to have anti-osteoarthritic effects, the potential bioactivity of Ryupunghwan (RPH), a traditional Korean medicinal botanical formula that contains *Astragalus membranaceus*, *Turnera diffusa*, *Achyranthes bidentata*, *Angelica gigas*, *Eclipta prostrata*, *Eucommia ulmoides*, and *Ilex paraguariensis*, is not known well. Therefore, the inhibitory effects of single compounds isolated from RPH on the OA-related molecules were investigated using IL-1β-stimulated chondrosarcoma SW1353 (SW1353) cell model. Two bioactive compounds, isomucronulatol 7-*O*-β-D-glucoside (IMG) and ecliptasaponin A (ES) were isolated and purified from RPH using column chromatography, and then the structures were analyzed using ESI-MS, ^{1}H-NMR, and ^{13}C-NMR spectrum. The expression or amount of matrix metalloproteinase 13 (MMP13), COX1/2, TNF-α, IL-1β or p65 was determined by RT-PCR, Western blot, and enzyme-linked immunosorbent assay (ELISA). RPH pretreatment reduced the expression and amounts of MMP13, and the expression of collagen II, COX1/2, TNF-α, IL-1β or p65, which were increased in IL-1β-stimulated SW1353 cells. IMG reduced the expression of all OA-related molecules, but the observed inhibitory effect was less than that of RPH extract. The other single compound ES showed the reduced expression of all OA-related molecules, and the effect was stronger than that in IMG (approximately 100 fold). Combination pretreatment of both single components remarkably reduced the expression of MMP13, compared to each single component. These synergic effects may provide potential molecular modes of action for the anti-osteoarthritic effects of RPH observed in clinical and animal studies.

Keywords: chondrosarcoma cells; osteoarthritis; Ryupunghwan (natural product mixture); IL-1β; isomucronulatol 7-*O*-β-D-glucoside; ecliptasaponin A

1. Introduction

The natural product formula popular in Asian countries, Ryupunghwan (RPH), is expected to provide pain relief and reduce inflammation as a health supplement. RPH *contains Astragalus membranaceus, Turnera diffusa, Achyranthes bidentata, Angelica gigas, Eclipta prostrata, Eucommia ulmoides,* and *Ilex paraguariensis*.

Astragalus membranaceus contains calycosin and is a known Asian medicinal herb traditionally used for the treatment of several diseases, such as hypertension, cirrhosis or cancer therapy through anti-inflammatory and anti-carcinogenic properties, respectively [1]. *Turnera diffusa*, which contains a major constituent arbutin, possesses anti-ulcer activity, which could be attributed to the inhibition of lipid peroxidation, immunomodulatory and anti-oxidant activities [2]. *Achyranthes bidentata* has been shown to protect rat articular chondrocytes against interleukin-1β-induced inflammation and apoptosis in vitro [3]. Thus, it is suggested that it might be a potential drug candidate in the treatment of osteoarthritis (OA) [3]. *Angelica gigas*, known as Chinese Angelica, has been shown to prevent diabetes and liver disease [4]. *Eclipta prostrata* has anti-inflammatory activity in a murine model of asthma [5]. *Eucommia ulmoides* ameliorates arthritis through inhibition of pro-inflammatory cytokines, and through reducing the degradation of cartilage and bone in rat collagen-induced arthritis [6]. *Ilex paraguariensis* can modulate antioxidant defense during perimenopause [7]. Thus, RPH was made by various natural products that have anti-inflammatory effects and reduce the degradation of cartilage, as suggested by folk remedies or laboratory experiments.

OA, which affects most mammalian populations, is the most common form of arthritis and is one of the leading causes of disability worldwide. In humans, 9.6% of men and 18% of women over the age of 60 years have symptomatic OA [8]. Ageing, obesity, gender, increased biomechanical loading of joints, and genetics or low-grade systemic inflammation have been known as risk factors for OA [9,10]. Although the mechanistic details of OA pathogenesis remain to be elucidated, the cartilage degradation and inflammation is the most predominant pathological feature that inevitably leads to joint dysfunction [11,12]. The articular cartilage is composed of water and extracellular matrix (ECM), which mainly composed of Type II collagen, aggrecan, proteoglycans and other collagen subtypes [10]. Thus, the cartilage homeostasis is maintained by the balance between ECM synthesis and degradation [12]. Disturbances of this balance are one of the main characteristics of OA cartilage, and the restoration of balance is the key factor for cartilage regeneration in OA. Inflammation in early OA is caused through inflammatory mediators, particularly TNF-α, IL-1β [13], and prostaglandin E2 (PGE_2), which is produced through degradation of arachidonic acid by COX-2 [14].

Recently, NSAIDs, which are used in the clinic for OA therapy, target the cyclooxygenase (COX) enzymes, key enzymes in the synthesis of prostaglandins produced in sites of tissue damage or infection, and inhibit both enzymes COX-1 and COX-2. Inhibition of COX-1 often results in gastrointestinal adverse effects, and the COX-2 is known as a regulator of inflammation [15]. The selective COX-2 inhibitor like celecoxib is used effectively for the pain relief and inflammation of OA [16,17]. Recently, new biological agents, such as bone morphogenetic protein-7 and growth factors etc., are known to stimulate chondrogenesis, inhibit matrix degradation and reduce inflammation [18]. However, in addition to an increased side effect such as gastrointestinal, cardiovascular adverse events and complications [19], these agents have failed to block the progression of OA [18,20]. Therefore, safer and better-tolerated solutions for the treatment of OA need to be developed.

Based on the findings above, this study investigates for the first time whether the RPH has inhibitory effects on the OA-related molecules in IL-1β-stimulated SW1353 cells. After confirmation the inhibitory effects of OA, this study also aimed to isolate single functional components contained in RPH and to investigate whether the single components have osteoarthritic health benefits. The data suggest that RPH may contribute to the development of a health functional food and valuable information for OA.

2. Results

2.1. *Effects of Natural Product Mixture (Ryupunghwan, RPH) on the Expression of Osteoarthritis (OA)-Related Molecules or Cyclooxygenase (COX) in IL-1β-Stimulated Chondrosarcoma SW1353 Cells*

The OA joints when stimulated by IL-1β, produce high levels of a variety of MMPs, particularly MMP13 [21,22]. Thus, we examined whether RPH has any influence on the expression of various molecules (MMP13, TNF-α, IL-1β, IκBα, and COX1/2) related to cartilage degradation and inflammation of OA in IL-1β-stimulated SW1353 cells. Expression of MMP13 mRNA and protein, and the amount of MMP13 secreted was increased in IL-1β-stimulated SW1353 cells versus negative control (NC). RPH pretreatment (50, 100, 300 μg/mL) reduced the expression of MMP13 in a dose-dependent manner (Figure 1A).

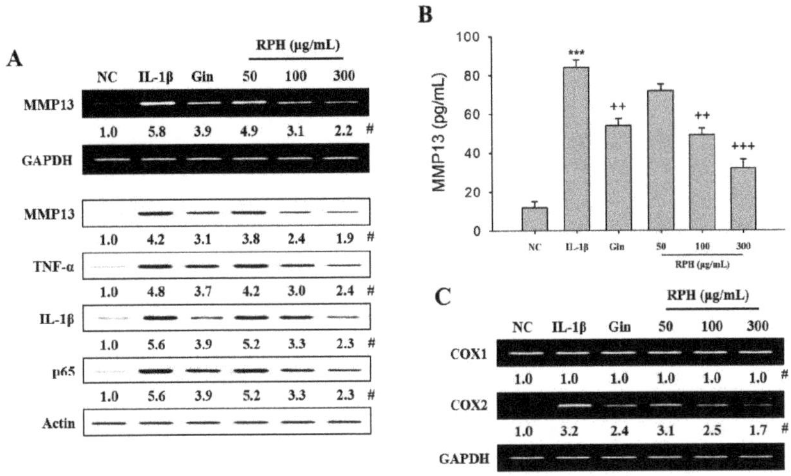

Figure 1. Effects of natural product mixture (Ryupunghwan (RPH)) on the expression or amount of osteoarthritis (OA)-related molecules or cyclooxygenase (COX) in IL-1β-stimulated chondrosarcoma SW1353 cells (SW1353 cells). SW1353 cells (1×10^6 cells) were stimulated with 20 μg/mL IL-1β for 24 h. Natural product mixture (RPH, 50, 100, 300 μg/mL) or ginsenoside (Gin, 300 mg/mL) was pretreated for 1 h before IL-1β. The expression of MMP13, COX-1, COX-2, TNF-α or IκBα was determined in protein or mRNA extracts isolated from the cell lysates using Reverse Transcription-Polymerase Chain Reaction (RT-PCR) or Western blot, respectively. Amount of MMP13 was determined in the cell supernatant isolated from IL-1β-stimulated cells using enzyme-linked immunosorbent assay (ELISA) kit. (**A**) Expression of MMP13 and OA-related molecules (TNF-α, IL-1β, or IκBα). (**B**) Amount of MMP13. (**C**) Expression of COX1/2. #, Numbers below band images show the mean values (n = 4) obtained from the ratio of band density of each group versus those of the control and loading control GAPDH or actin. Results for ELISA assay represent the mean ± SEM (n = 4) obtained from four independent experiments performed in triplicates. NC, negative control; IL-1β (interleukin 1 beta), IL-1β-stimulated SW1353 cells; Gin, ginsenoside. ***, $p < 0.001$ versus the NC. $^{++}$, $p < 0.01$; $^{+++}$, $p < 0.001$ versus the IL-1β stimulation.

It has been reported that TNF-α is pro-inflammatory cytokine involved in IL-1β-induced chondrocytes or OA [13,23], that NF-κB is related to the production of inflammatory molecules in chondrocytes [24,25]. Thus, we examined whether RPH affects the expression of OA related-molecules in IL-1β-stimulated SW1353 cells. The expression of TNF-α, IL-1β, and NF- kB subunit p65 was significantly increased in IL-1β-stimulated SW1353 cells versus NC. RPH pretreatment inhibited these molecules, which are typically increased in IL-1β-stimulated SW1353 cells, in the dose-dependent manner (Figure 1A), but it increased expression of IκBα. Pretreatment of 300 μg/mL (high dose of

RPH) of ginsenoside, which is used as one of natural positive control of natural product mixture (RPH), inhibited the expression of all OA-related molecules (MMP13, TNF-α, IL-1β) which were observed in RPH as well as amount of MMP13, but it increased expression of IκBα. However, its efficacy was less than those in 100 μg/mL of RPH (middle dose).

We examined whether RPH has an effect on the expression of COX1/2, which are known as a regulator of inflammation [21]. The expression of COX-2 was significantly increased in IL-1β-stimulated SW1353 cells versus NC. RPH pretreatment inhibited the increased expression of this enzyme in a dose-dependent manner. Pretreatment of 300 μg/mL Gin inhibited COX-2 expression less than that in 100 μg/mL RPH (middle dose). However, constitutive COX-1 did not show any change in this experimental groups pretreated with RPH or Gin (Figure 1C).

2.2. *Purification and Identification of Bioactive Ingredients from RPH*

To investigate single component, which has anti-OA efficacy contained in RPH, we used a variety of methods. Ryupunghwan (RPH) as a natural product mixture and its contents are described in the materials and methods section. RPH was suspended in distilled water under ultrasonic agitation at 90 Hz and 40 °C and successively partitioned with ethyl acetate and n-BuOH to afford ethyl acetate and n-BuOH, and water fractions. n-BuOH extracts of RPH showed potent inhibitory effects on the expression of MMP13 (data not shown). n-BuOH fraction was subjected to silica gel and C-18 column chromatography, and five compounds (**1**–**5**) were isolated (Figure 2). The spectroscopic data (ESI-MS, ^{1}H-NMR, and ^{13}C-NMR spectra) and comparisons with previous data confirmed that these structures were (−)-pinoresinol 4-*O*-β-D-glucopyranoside (**1**) [26], (−)-marmesinin (**2**) [27], columbianetin β-D-glucopyranoside (**3**) [28], isomucronulatol 7-*O*-β-D-glucoside (**4**) [29], and ecliptasaponin A (**5**) [30] (Figure 2, Supplementary Materials Figures S1–S15).

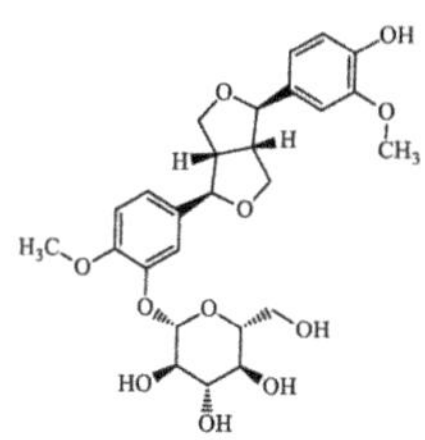

1

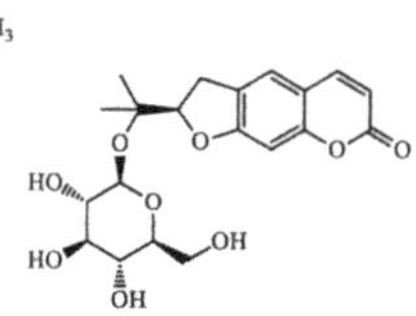

2

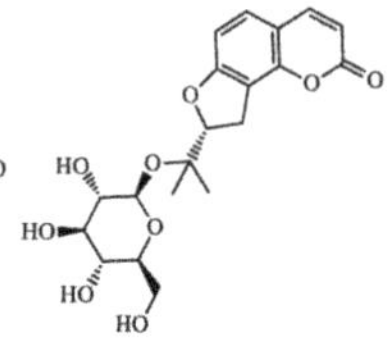

3

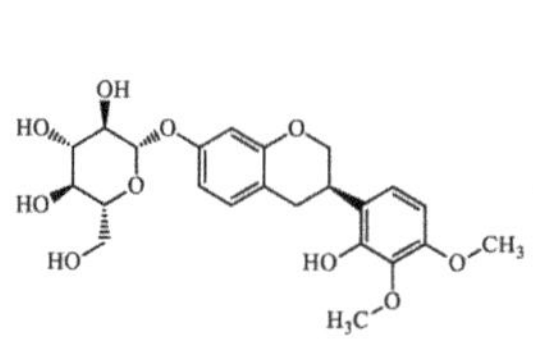

4

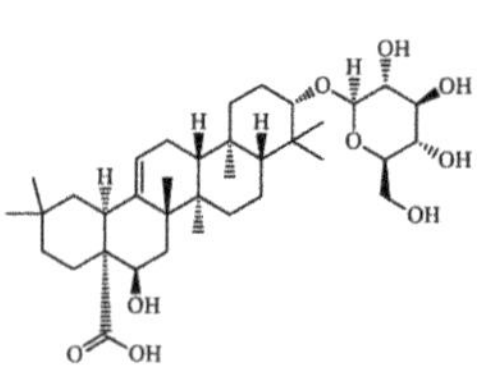

5

Figure 2. Chemical structures of compounds isolated from RPH. RPH (1.5 kg) was fractioned using various organic chemicals and columns as described in "Materials and Methods". Finally, five single components were identified from RPH, and then structure of each component was analyzed using ESI-MS, ^{1}H-NMR, and ^{13}C-NMR spectrum. Inhibitory effects of each fraction identified from organic chemicals were determined in the expression of MMP13 (data not shown). Structures of (**1**), (**2**), (**3**), (**4**), or (**5**) indicate (−)-pinoresinol 4-*O*-β-D-glucopyranoside, (−)-marmesinin, columbianetin β-D-glucopyranoside, isomucronulatol 7-*O*-β-D-glucoside (IMG) and ecliptasaponin A (ES), respectively.

2.3. Effects of Single Component IMG and ES Isolated from Natural Product Mixture (RPH) on the Expression of OA-Related Molecules in IL-1β-Stimulated SW1353 Cells

Five single compounds were isolated from RPH, and to be the responsible bioactive compounds. Two substances (isomucronulatol 7-O-β-D-glucoside (IMG) and ecliptasaponin A (ES)) among five substances showed the suppression of molecules, MMP13, collagen type II, TNF-α, IL-1β, and COX-2, typically related to OA. One of single components IMG (30, 50, 100 μg/mL) suppressed the expression of MMP13, collagen type II, TNF-α, IL-1β, and COX-2 related to OA in IL-1β-stimulated SW1353 cells in a dose-dependent manner (Figure 3). The other single component ES (10, 30, 50 ng/mL) remarkably reduced the expression of all molecules related to OA versus IL-1β-stimulated SW1353 cells. IMG showed less potency than that in RPH. On the other hand, ES has stronger potency than that of RPH or IMG, but toxicity in terms of cell lysis was observed in the high dose above 70 ng/mL (data not shown). However, the potency of ES was stronger by approximately 1000 times than IMG. Therefore, we tried the combination pretreatment of both single functional components. The combination pretreatment of IMG (30, 50, 100 μg/mL) with ES, which fixed on 50 ng/mL of dose, remarkably suppressed expression of all OA-related molecules by approximately 48~67% (100 μg/mL of IMG plus 50 ng/mL of ES) in IL-1β-stimulated SW1353 cells (Figure 3).

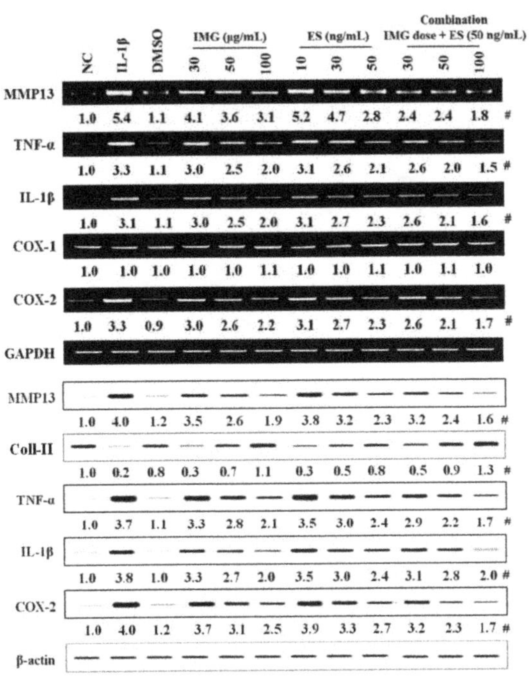

Figure 3. Effects of single component IMG and ES isolated from natural product mixture (RPH) on the expression of OA-related molecules in IL-1β-stimulated SW1353 cells. SW1353 cells (1×10^6 cells) were stimulated with 20 μg/mL IL-1β for 24 h. IMG (isomucronulatol 7-β-O-glucoside, 30, 50 or 100 μg/mL), ES (ecliptasaponin A, 10, 30 or 50 ng/mL), or combination (each dose of IMG + 50 ng/mL ES) was pretreated at 1 h before IL-1β stimulation. The expression of OA-related molecules including collagen type II was determined in mRNA extracts isolated the cell lysates using RT-PCR, and Western blot, respectively. (Upper panel) The mRNA expression of OA-related molecules (MMP13, TNF-α, IL-1β, COX1/2). (Lower panel) The protein expression of OA-related molecules (MMP13, Collagen type II, TNF-α, IL-1β, and COX-2). #, Numbers below band images are the mean values (n = 4) obtained from as described in the Figure 1 legend. NC, negative control; IL-1β (interleukin 1 beta), IL-1β-stimulated SW1353 cells; DMSO (0.3%), control for the solvent of a single component; Coll-II, collagen type II.

3. Discussion

We demonstrate that natural product mixture Ryupunghwan (RPH) reduces the expression or secretion of OA-related molecules such as MMP13, TNF-α, p65 and COX1/2, and that two compounds (Isomucronulatol 7-*O*-β-D-glucoside, IMG; Ecliptasaponin A, ES) among five single compounds, which are isolated and purified from RPH and then their structures are analyzed, have functional activities. A single functional compound IMG is a known compound present in *Astragalus propinquus* [1,31] that has the highest percentage in RPH, but its functional activity for OA has not been evaluated yet. *Astragalus propinquus* contains calycosin-7-*O*-β-D-glucopyranoside as an index material, which had anti-inflammation and anti-osteoarthritis properties [32,33]. The functional activity in OA for ES was also not clarified yet, although ES is found in *Eclipta prostrata* L. and is reported as an index material, which exhibited anti-inflammatory activity in an allergic asthma model in mice [5]. There is only one report that ES shows a protective effect against lung tissue inflammation in the bleomycin-induced pulmonary fibrosis via reducing the oxidative stress [34].

The pathological feature for OA, which is caused by various risk factors such as ageing and low-grade systemic inflammation etc., is characterized by cartilage degradation and inflammation of joint [11,13]. Increased MMP13 expression is related to cartilage degradation [21,22]. Inflammatory cytokines TNF-α and IL-1β are associated with inflammation of joint in early OA [13,14]. NF-κB subunit p65 as a transcriptional factor causes the production of inflammatory cytokines [14,24,25]. The articular cartilage is mainly composed of collagen type II, aggrecan and proteoglycans subtypes [10]. Prostaglandins (PGE_2), which recruited/activated inflammatory cells into synovia, are produced via COX2 [14]. Thus, our data suggest that RPH may reduce cartilage degradation and inflammation in OA through inhibiting the expression of MMP13, collagen type II, COX2 and inflammatory cytokines, as demonstrated by the data showing that RPH reduced expression of all OA-related molecules investigated. The data can also be inferred that RPH may induce the regeneration of cartilage via reducing the expression of MMP13 and collagen type II in OA.

RPH has no effect on COX1 expression, which is known as an enzyme to regulate the secretion of gastric acid in gastric intestinal tract [35]. This phenomenon implies that RPH is natural product mixture that could be used for OA management and have less toxicity because it has no influence on COX1 expression.

Also, our observations suggest that single component IMG and ES isolated in RPH may inhibit the expression of MMP13 and COX2, which is typically associated with the degeneration of cartilage in osteoarthritis, and inflammatory cytokines, and expression of COX2 which is known as a regulator of inflammation.

In conclusion, these in vitro results support that single functional compounds IMG and ES have significantly anti-osteoarthritic effect in IL-1β-stimulated chondrosarcoma SW1353 cells. The data suggest that RPH containing IMG and ES with less side-effects, which have anti-oxidative and anti-inflammatory properties associated with osteoarthritis, may have a potential as a health functional food supplement for osteoarthritis, with less side-effects and should be further evaluated in animal and clinical models.

4. Materials and Methods

4.1. Materials

Ryupunghwan (RPH), a natural product mixture was donated by Hyunsung Vital Co. Ltd. (Seoul, Korea). RPH contains *Astragalus propinquus* (31%), *Turnera diffusa* (14%), *Achyranthes bidentate* (14%), *Angelica sinensis* (14%), *Eclipta prostrata* (12%), *Eucommia ulmoides* (8%), and *Ilex paraguariensis* (7%). These plants were extracted using hot water (90 °C) for 24 h and evaporated by Liquefied extractor (Hyunsung Vital Co. Ltd., Seoul, Korea) to yield a powder of RPH.

4.2. Purification and Identification of Bioactive Ingredients

The Ryupunghwan (RPH, 1.5 kg) was suspended in distilled water (4.0 L) under ultrasonic agitation at 90 Hz and 40 °C and successively partitioned with ethyl acetate and n-BuOH to afford ethyl acetate (96.0 g, A) and n-BuOH (80.0 g, B), and water fractions.

The n-BuOH fraction (B) showed potent inhibitory effects on the expression of osteoarthritis (OA)-related molecules or cyclooxygenase (COX), this fraction was chosen for subsequent studies. The n-BuOH fraction was separated using a silica gel column with a gradient solvent mixture of $CHCl_3$-MeOH-H_2O (10:1:0, 6:1:0.1, 4:1:0.1, 2:1:0.1, and 1:1:0.1) to afford six subfractions (B-1 to B-6). Next, subfraction B-1 (12.0 g) was subjected to silica gel CC and was eluted with a solvent mixture of $CHCl_3$-acetone (10:1, and 7:1) and CHCl3-MeOH-H_2O (7:1:0.1, 5:1:0.1) to afford four subfractions (B-1.1 to B-1.4). Further purification of subfraction B-1.1 (1.4 g) via YMC RP-C18 silica gel column using mixtures of MeOH–H_2O (1:1, 1.5:1, and 2:1) yielded compounds **1** (80.0 mg), **2** (78.0 mg), and **3** (90.0 mg). When the same steps were repeated as above, compounds **4** (20.0 mg), and **5** (60.0 mg) were also obtained by purifying subfraction B-1.2 (2.0 g) on YMC RP-C18 silica gel using mixtures of MeOH–H_2O (1.5:1, and 2:1).

4.3. Human Chondrosarcoma SW1353 Cells Culture Conditions

Human chondrosarcoma cells (SW1353 cells) were obtained from American Type Culture Collection (ATCC; No. HTB-94, Manassas, VA, USA). The cells were grown in Leibovitz's L-15 medium (Gibco, Grand Island, NY, USA) supplemented with 1% l-glutamine (Sigma-Aldrich, St. Louis, MO, USA), 1% antibiotic penicillin/streptomycin solution (Sigma-Aldrich, St. Louis, MO, USA), and 10% fetal bovine serum (HyClone, Logan, UT, USA). The cells were maintained at 37 °C in a humidified atmosphere without CO_2, and the media was replaced every 3 days [36].

4.4. Cell Stimulation and Treatment

After serum starvation for 24 h, the cells (1×10^6 cells) were stimulated with 20 μg/mL recombinant human interleukin-1 beta (IL-1β; PeproTech, Rocky Hill, NJ, USA), and then incubated for 24 h [37]. The cells were centrifuged (470× *g*, 3 min) to separate the supernatants and cells. The supernatants were used to measure the amount of MMP13, and the cells used to determine the expression of all molecules related to OA. Optimal concentrations for IL-1β stimulation, RPH, single component or Gin (20 μg/mL) were yielded in the preliminary experiments (data not shown). Dimethyl sulfoxide (DMSO; 0.3%), ginsenoside (300 μg/mL), natural product mixture (RPH; 50, 100 or 300 μg/mL), or isomucronulatol 7-β-*O*-glucoside (SG; 30, 50 or 100 μg/mL) and ecliptasaponin A (ES; 10, 30 or 50 ng/mL) as a single component separated from natural mixture were pretreated at 1h before IL-1β stimulation.

4.5. Reverse Transcription-Polymerase Chain Reaction (RT-PCR)

Total mRNA was extracted from the SW1353 cells (1×10^6 cells) using TRIzol reagent (Invitrogen, Life Technologies Ltd., UK). RT-PCR was performed in a final volume of 20 μL using a High-capacity cDNA Reverse Transcription kit (Applied Biosystems, Foster City, CA, USA) and G-taq kit (Cosmogenetech, Seoul, Korea) in an automated thermal cycler (Bio-Rad, Laboratories, CA, USA). PCR assays were performed for 35 cycles. Each cycle consisted of following steps: denaturation at 94 °C for 30 s, annealing at 56 °C for 45 s, and extension at 72 °C for 1 min. The result was expressed as a ratio of GAPDH mRNA. PCR products were analyzed using 1% agarose gel and visualized under a UV light after staining with stay safe nucleic acid gel stain (Real Biotech Corporation, Banqiao, Taiwan) [38].

The primer sequences used were as follows: MMP13 sense, 5′-TCC CAG GAA TTG GTG ATA AAG TAG A-3′; MMP13 anti-sense, 5′-CTG GCA TGA CGC GAA CAA TA-3′; TNF-α sense, 5′-TCT ACT CCC AGG TCC TCT TC-3′; TNF-α anti-sense, 5′-AAG TAG ACC TGC CCA GAC TC-3′; IL-1β sense, 5′-CTT TGA AGC TGA TGG CCC TAA A-3′; IL-1β anti-sense, AGT GGT GGT CGG AGA TTC GTA-3′; COX1 sense, 5′-CGC GGA TCC ACC ATG AGC CGG-3′; COX1 anti-sense, 5′-TGC TTT CAA GCT TCT CAG-3′; COX2 sense, 5′-TTG CGG CCG CCA CCA TGG TCG -3′; COX2 anti-sense, 5′-GCT CTA GAG ACT TCT ACA GTT CAG-3′; GAPDH sense, 5′-AAC TTT GGC ATT GTG GAA GG-3′; GAPDH anti-sense, 5′-ACA CAT TGG GGG TAG GAA CA-3′.

4.6. Preparation of Nuclear Extracts

The SW1353 cells (1 × 10^6 cells) harvested from IL-1β-stimulated cells were suspended in a cytoplasmic extraction buffer [10 mM HEPES, 60 mM KCl, 1 mM EDTA, 0.075% (*v*/*v*) NP40, 1mM DTT, 1 mM PMSF and 2.5 ug/mL each of aprotinin, leupeptin, and pepstatin), adjusted to pH 7.6], and allowed to incubate on ice for 15 min. After removing the cytoplasmic extract, the cell pellets were washed with cytoplasmic extraction buffer without NP40. After spin down, and the cell pellet was treated with the nuclear extraction buffer [20 mM Tris Cl, 420 mM NaCl, 1.5 mM $MgCl_2$, 0.2 mM EDTA, 0.5 mM PMSF and 25% (*v*/*v*) glycerol, adjusted to pH 8.0]. The final salt concentration was adjusted to 400 mM with NaCl. The extracts were incubated on ice for 30 min, and then the supernatants were collected [39]. Nuclear extracts (50 μg) in supernatants were used to measure for expression p65 using Western blot.

4.7. Western Blot Analysis

The SW1353 cells (1 × 10^6 cells) harvested from IL-1β-stimulated cells were suspended in a low-salt lysis buffer [10 mM HEPES (pH 7.9), 10 mM KCl, 0.1 mM EDTA, 0.1 mM EGTA, 1 mM DTT, 0.5 mM PMSF, 2 μg/mL aprotinin, 2 μg/mL leupeptin] and allowed to swell on ice for 30 min. The cells were then homogenized using a Polytron homogenizer (Kinematica, Lucern, Switzerland). After centrifugation, supernatants obtained from cells extracts were analyzed by 10% SDS-polyacrylamide gel electrophoresis and electrophoretically transferred to nitrocellulose membranes (Amersham Biosciences, Piscataway, NJ, USA). The membranes were washed with PBS containing 0.1% Tween 20 (PBST) and then blocked for 1 h in PBST containing 5% skim milk. After washing the membranes with PBST, they were treated with primary Abs against actin, MMP13, TNF-α, IL-1β, COX-2, or P65 (Cell Signaling Technology, Bevery, MA, USA), or collagen type II (Santa Cruz, CA, USA) diluted with PBST (1:1000). Membranes were washed with PBST and treated with horseradish peroxidase (HRP)-conjugated goat anti-mouse or HRP-conjugated goat anti-rabbit IgG (diluted to 1:5000~1:10,000) (Bethyl Laboratories, Montgomery, TX, USA) in PBST for 1 h. After washing, the protein bands were visualized by Enhanced Chemi-Luminescence using a chemiluminometer (ECL; Amersham Biosciences, Piscataway, NJ, USA) [40].

4.8. MMP13 Amount Assay

The amount of MMP13 secreted in supernatants (100 μL) isolated from media of cells stimulated with IL-1β was determined using MMP13 ELISA kit (Abcam, Cambridge, UK). Briefly, 100 μL standard solution or samples were added in 96 well plates coated with specific human MMP13 and then incubated for an overnight on a microplate shaker. Each well was washed 3 times with 300 μL of diluted wash buffer, and then 150 μL of HRP-conjugated streptavidin (Abcam, Cambridge, UK) was added and incubated for 1 h. The wells are again washed, a TMB substrate solution was added to the wells, and the color was developed in proportion to a number of MMP13 bound. The plates were read at 450 nm using microplate spectrophotometer (Molecular Devices, Sunnyvale, CA, USA). The amounts of MMP13 were calculated using the standard curves generated by specific MMP13 standards. The results were expressed in ng/mL (1 × 10^6 cells). The lowest detection limit for MMP13 was 8.23 pg/mL [41].

4.9. Statistical Analysis

Experimental data are shown as means ± SEM (n = 8). The unpaired Student's t-test was used to compare the two groups. Multiple group comparisons were performed using two-way analysis of variance (ANOVA) followed by Scheffe's post hoc test, using the SPSS 11 software (SPSS Inc., Chicago, IL, USA). p values < 0.05 were considered to indicate statistical significance. The densitometry analysis of Western blots and RP-PCR were performed with Quantity One (version 4.6.3; Bio-Rad, Hercules, CA, USA) and are indicated as means ± SEM (n = 4) obtained from the ratio of each band density versus those in the control and loading control of four independent experiments using the densitometer.

Supplementary Materials: The following are available online. Figure S1: ^{1}H-NMR of (−)-pinoresinol 4-O-β-D-glucopyranoside (**1**); Figure S2: ^{13}C-NMR of (−)-pinoresinol 4-O-β-D-glucopyranoside (**1**); Figure S3: ESI-MS spectra of (−)-pinoresinol 4-O-β-D-glucopyranoside (**1**); Figure S4: ^{1}H-NMR spectra of (−)-marmesinin (**2**); Figure S5: ^{13}C-NMR spectra of (−)-marmesinin (**2**); Figure S6: ESI-MS spectra of (−)-marmesinin (**2**); Figure S7: ^{1}H-NMR spectra of columbianetin β-D-glucopyranoside (**3**); Figure S8: ^{13}C-NMR spectra of columbianetin β-D-glucopyranoside (**3**); Figure S9: ESI-MS spectra of columbianetin β-D-glucopyranoside (**3**); Figure S10: ^{1}H-NMR spectra of isomucronulatol 7-O-β-D-glucoside (**4**); Figure S11: ^{13}C-NMR spectra of isomucronulatol 7-O-β-D-glucoside (**4**); Figure S12: ESI-MS spectra of isomucronulatol 7-O-β-D-glucoside (**4**); Figure S13: ^{1}H-NMR spectra of ecliptasaponin A (**5**); Figure S14: ^{13}C-NMR spectra of ecliptasaponin A (**5**); Figure S15: ESI-MS spectra of ecliptasaponin A (**5**).

Author Contributions: J.Y.R. designed the study and directed its implementation, including the study's analytic strategy. Y.-I.K. helped conduct the literature review and prepare the Methods and the Discussion sections of the text. Y.H.K., H.-B.C., J.-Y.L., H.K. and T.Y.K. prepared the product to be tested and participated in preparation of the manuscript. G.U.H., E.Y.H., J.Y.P., S.H.J. and Y.H.S. conducted the animal cell experiment and analyzed the data. All the authors read and approved the final manuscript.

Funding: This study was supported by a research grant from Hannam University in 2018.

Conflicts of Interest: The authors declare no conflict of interest.

References

1. Gao, J.; Liu, Z.J.; Chen, T.; Zhao, D. Pharmaceutical properties of calycosin, the major bioactive isoflavonoid in the dry root extract of Radix astragali. *Pharm. Biol.* **2014**, *52*, 1217–1222. [CrossRef] [PubMed]
2. Taha, M.M.; Salga, M.S.; Ali, H.M.; Abdulla, M.A.; Abdelwahab, S.I.; Hadi, A.H. Gastroprotective activities of Turnera diffusa Willd. ex Schult. revisited: Role of arbutin. *J. Ethnopharmacol.* **2012**, *141*, 273–281. [CrossRef] [PubMed]
3. Xu, X.X.; Zhang, X.H.; Diao, Y.; Huang, Y.X. Achyranthes bidentate saponins protect rat articular chondrocytes against interleukin-1β-induced inflammation and apoptosis in vitro. *Kaohsiung. J. Med. Sci.* **2017**, *33*, 62–68. [CrossRef] [PubMed]
4. Wang, K.; Tang, Z.; Zheng, Z.; Cao, P.; Shui, W.; Li, Q.; Zhang, Y. Protective effects of Angelica sinensis polysaccharide against hyperglycemia and liver injury in multiple low-dose streptozotocin-induced type 2 diabetic BALB/c mice. *Food Funct.* **2016**, *7*, 4889–4897. [CrossRef] [PubMed]
5. Morel, L.J.; Azevedo, B.C.; Carmona, F.; Contini, S.H.; Teles, A.M.; Ramalho, F.S.; Bertoni, B.W.; de Castro França, S.; de Carvalho Borges, M.; Pereira, A.M. A standardized methanol extract of *Eclipta prostrata* (L.) L. (Asteraceae) reduces bronchial hyperresponsiveness and production of Th2 cytokines in a murine model of asthma. *J. Ethnopharmacol.* **2017**, *198*, 226–234. [CrossRef] [PubMed]
6. Wang, J.Y.; Yuan, Y.; Chen, X.J.; Fu, S.G.; Zhang, L.; Hong, Y.L.; You, S.F.; Yang, Y.Q. Extract from *Eucommia ulmoides* Oliv. ameliorates arthritis via regulation of inflammation, synoviocyte proliferation and osteoclastogenesis in vitro and in vivo. *J. Ethnopharmacol.* **2016**, *194*, 609–616. [PubMed]
7. Pereira, A.A.; Tirapeli, K.G.; Neto, A.H.; da Silva Brasilino, M.; da Rocha, C.Q.; Belló-Klein, A.; Llesuy, S.F.; Dornelles, R.C.; de Melo Stevanato Nakamune, A.C. Ilex paraguariensis supplementation may be an effective nutritional approach to modulate oxidative stress during perimenopause. *Exp. Gerontol.* **2017**, *90*, 14–18. [CrossRef] [PubMed]
8. Palmieri, B.; Lodi, D.; Capone, S. Osteoarthritis and degenerative joint disease: Local treatment options update. *Acta Biomed.* **2010**, *81*, 94–100. [PubMed]

9. Martel-Pelletier, J.; Barr, A.J.; Cicuttini, F.M.; Conaghan, P.G.; Cooper, C.; Goldring, M.B.; Golgring, S.R.; Jones, G.; Teichtahl, A.J.; Pelletier, J.P. Osteoarthritis. *Nat. Rev. Dis. Primers* **2016**, *2*, 16072. [CrossRef] [PubMed]
10. Varela-Eirin, M.; Loureiro, J.; Fonseca, E.; Corrochano, S.; Caeiro, J.R.; Collado, M.; Mayan, M.D. Cartilage regernation and ageing: Targeting cellular plasticity in osteoarthritis. *Ageing Res. Rev.* **2018**, *42*, 56–71. [CrossRef] [PubMed]
11. Van den Berg, W.B. Osteoarthritis year 2010 in review: Pathomechanism. *Osteoarthritis Cartilage* **2011**, *19*, 338–341. [CrossRef] [PubMed]
12. Sun, H.; Wu, Y.; Pan, Z.; Yu, D.; Chen, P.; Zhang, X.; Wu, H.; Zhang, X.; An, C.; Chen, Y.; et al. Gefitinib for epidermal growth factor receptor activated osteoarthritis subpopulation treatment. *EBioMedicine* **2018**, *32*, 223–233. [CrossRef] [PubMed]
13. Boehme, K.A.; Rolauffs, B. Onset and progression of human osteoarthritis-can growth factors, inflammatory cytokines, or differential miRNA expression concomitantly induce proliferation, ECM degradation, and inflammation in articular cartilage? *Int. Mol. Sci.* **2018**, *19*, 2282. [CrossRef] [PubMed]
14. Liu, C.C.; Zhang, Y.; Dai, B.-L.; Ma, Y.-J.; Zhang, Q.; Wang, Y.; Yang, H. Chlorogenic acid prevents inflammatory responses in IL-1β-stimulated human SW-1353 chondrocytes, a model for osteoarthritis. *Mol. Med. Rep.* **2017**, *16*, 1369–1375. [CrossRef] [PubMed]
15. Yan, L.; Pan, M.; Fu, M.; Wang, J.; Huang, W.; Qian, H. Design, Synthesis and biological evaluation of novel analgesic agents targeting both cyclooxygenase and TRPV1. *Bioorg. Med. Chem.* **2016**, *24*, 849–857. [CrossRef] [PubMed]
16. Essex, M.; Bhadra, P.; Sands, G. Efficacy and tolerability of celecoxib versus naproxen in patients with osteoarthritis of the knee: A randomized, double-blind, double-dummy trial. *J. Int. Med. Res.* **2012**, *40*, 1357–1370. [CrossRef] [PubMed]
17. Zhu, X.; Wu, D.; Sang, L.; Wang, Y.; Shen, Y.; Zhuang, X.; Chu, M.; Jiang, L. Comparative effectiveness of glucosamine, chondtoitin, acetaminophen or celecoxib for the treatment of kee and/or hip osteoarthritis: A network meta-analysis. *Clin. Exp. Rheumatol.* **2018**, *36*, 595–602. [PubMed]
18. Singh, R.; Akhtar, N.; Haqqi, T.M. Green tea polyphenol epigallocatechin-3-gallate: Inflammation and arthritis. *Life Sci.* **2010**, *86*, 907–918. [CrossRef] [PubMed]
19. Kalamgam, G.; Memic, A.; Budd, E.; Abbs, M.; Mobasheri, A. A comprehensive review of stem cells for cartilage regeneration in osteoarthritis. *Adv. Exp. Med. Biol.* **2018**. [CrossRef]
20. Chen, B.; Qin, J.; Wang, H.; Magdalou, J.; Chen, L. Effects of adenovirus-mediated bFGF, IL-1Ra and IGF-1gene transfer on human osteoarthritis chondrocytes and osteoarthritis in rabbits. *Exp. Mol. Med.* **2010**, *42*, 684–695. [CrossRef] [PubMed]
21. Ma, Z.; Wang, Y.; Piao, T.; Liu, J. Echinocystic Acid inhibits IL-1β-Induced COX-2 and iNOS expression in human osteoarthritis chondrocytes. *Inflammation* **2016**, *39*, 543–549. [CrossRef] [PubMed]
22. Ruan, G.; Xu, J.; Wang, K.; Wu, J.; Zhu, Q.; Ren, J.; Bian, F.; Chang, B.; Bai, X.; Han, W.; et al. Association between knee structural measures, circulating inflammatory factors and MMP13 in patients with knee osteoarthritis. *Osteoarthritis Cartilage* **2018**, *26*, 1063–1069. [CrossRef] [PubMed]
23. Bao, G.; Xu, L.; Xu, X.; Zhai, L.; Duan, C.; Xu, D.; Song, J.; Liu, Z.; Tao, R.; Cui, Z.; et al. SGTB promotes the caspase-dependent apoptosis in chondrocytes of osteoarthritis. *Inflammation* **2016**, *39*, 601–610. [CrossRef] [PubMed]
24. Wang, C.; Zeng, L.; Zhang, T.; Liu, J.; Wang, W. Tenuigenin prevents IL-1β-induced inflammation in human osteoarthritis chondrocytes by suppressing PI3K/AKT/NF-κB signaling pathway. *Inflammation* **2016**, *39*, 807–812. [CrossRef] [PubMed]
25. Wang, L.; Gai, P.; Xu, R.; Zheng, Y.; Lv, S.; Li, Y.; Liu, S. Shikonin protects chondrocytes from interleukin-1beta-induced apoptosis by regulating PI3K/Akt signaling pathway. *Int. J. Clin. Exp. Pathol.* **2015**, *8*, 298–308. [PubMed]
26. Ouyang, M.; Wein, Y.; Zhang, Z.K.; Kuo, Y.H. Inhibitory activity against tobacco mosaic virus (TMV) replication of pinoresinol and syringaresinol lignans and their glycosides from the root of *Rhus javanica* var. roxburghiana. *J. Agric. Food Chem.* **2007**, *55*, 6460–6465. [CrossRef] [PubMed]
27. Kwon, Y.S.; Woo, E.R.; Kim, C.M. A study on the constituents of bioactive fractions of *Ostericum koreanum* Kitagawa. *Korean J. Pharmacogn.* **1991**, *22*, 156–161.

28. Kim, Y.A.; Lee, J.I.; Kong, C.-S.; Choe, J.C.; Oh, K.S.; Seo, Y. Antioxidant activity of dihydrofurocoumarins from *Corydalis heterocarpa*. *Biotech. Bioproc. Eng.* **2014**, *19*, 771–779. [CrossRef]
29. Ma, X.; Tu, P.F.; Chen, Y.J.; Zhang, T.Y.; Wei, Y.; Ito, Y.I. Preparative isolation and purification of isoflavan and pterocarpan glycosides from *Astragalus membranaceus* Bge. var. *mongholicus* (Bge.) Hsiao by high-speed counter-current chromatography. *J. Chromatograph. A* **2004**, *1023*, 311–315. [CrossRef]
30. Yahara, S.J.; Ding, N.; Nohara, T.H. Oleanane glycosides from *Eclipta alba*. *Chem. Pharmaceu. Bull.* **1994**, *42*, 1336–1338. [CrossRef]
31. Kim, G.S.; Lee, D.Y.; Lee, S.E.; Noh, H.J.; Choi, J.H.; Park, C.G.; Choi, S.I.; Hong, S.J.; Kim, S.Y. Evaluation of extraction conditions and HPLC analysis method for bioactive compounds of Astragali Radix. *Korean J. Med. Crop. Sci.* **2013**, *21*, 486–492. [CrossRef]
32. Choi, S.I.; Park, S.R.; Heo, T.R. Inhibitory effect of Astragali Radix on matrix degradation in human articular cartilage. *J. Microbiol. Biotechnol.* **2005**, *15*, 1258–1266.
33. Choi, S.I.; Heo, T.R.; Min, B.H.; Cui, J.H.; Choi, B.H.; Park, S.R. Alleviation of osteoarthritis by calycosin-7-*O*-beta-*D*-glucopyranoside (CG) isolated from *Astragali Radix* (AR) in rabbit osteoarthritis (OA) model. *Osteoarthritis Cartilage* **2007**, *15*, 1086–1092. [CrossRef] [PubMed]
34. You, X.Y.; Xue, Q.; Fang, Y.; Liu, Q.; Zhang, C.F.; Zhao, C.; Zhang, M.; Xu, X.H. Preventive effects of Ecliptae Herba extract and its component, ecliptasaponin A, on bleomycin-induced pulmonary fibrosis in mice. *J. Ethnopharmacol.* **2015**, *175*, 172–180. [CrossRef] [PubMed]
35. Takeuchi, K.; Amagase, L. Roles of cyclooxygenase, prostaglandin E2 and EP receptors in mucosal protection and ulcer healing in the gastrointestinal tract. *Curr. Pharm. Des.* **2018**, *24*, 2002–2011. [CrossRef] [PubMed]
36. Chang, C.C.; Hsieh, M.S.; Liao, S.T.; Chen, Y.H.; Cheng, C.W.; Huang, P.T.; Lin, Y.F.; Chen, C.H. Hyaluronan regulates PPARγ and inflammatory responses in IL-1β-stimulated human chondrosarcoma cells, a model for osteoarthritis. *Carbohydr. Polym.* **2012**, *90*, 1168–1175. [CrossRef] [PubMed]
37. Wang, Z.; Ding, L.; Zhang, S.; Jiang, T.; Yang, Y.; Li, R. Effects of icariin on the regulation of the OPG-RANKL-RANK system are mediated through the MAPK pathways in IL-1β-stimulated human SW1353 chondrosarcoma cells. *Int. J. Mol. Med.* **2014**, *34*, 1720–1726. [CrossRef] [PubMed]
38. Hong, G.U.; Kim, N.G.; Kim, T.J.; Ro, J.Y. CD1d expressed in mast cell surface enhances IgE production in B cells by up-regulating CD40L expression and mediator release in allergic asthma in mice. *Cell. Signal.* **2014**, *26*, 1105–1117. [CrossRef] [PubMed]
39. Keifer, J.A.; Guttridge, D.C.; Ashburner, B.P.; Baldwin, A.S., Jr. Inhibition of NF-kappa B activity by thalidomide through suppression of IκappaB kinase activity. *J. Biol. Chem.* **2001**, *276*, 22382–22387. [CrossRef] [PubMed]
40. Ahn, Y.M.; Hong, G.U.; Kim, S.H.; Lee, H.J.; Baek, H.S.; Kim, M.N.; Park, K.Y.; Ro, J.Y. Transglutaminase 2 expressed in mast cells recruited into skin or bone marrow induces the development of pediatric mastocytosis. *Pediatr. Allergy Immunol.* **2015**, *26*, 438–445. [CrossRef] [PubMed]
41. Liao, S.; Zhou, K.; Li, D.; Xie, X.; Jun, F.; Wang, J. Schisantherin A suppresses interleukin-1β-induced inflammation in human chondrocytes via inhibition of NF-κB and MAPKs activation. *Eur. J. Pharmacol.* **2016**, *780*, 65–70. [CrossRef] [PubMed]

Sample Availability: Samples of the compounds are available from the corresponding authors (Y.I.K.; J.Y.R.).

molecules

MDPI

Article

Inhibitory Effects of Roseoside and Icariside E4 Isolated from a Natural Product Mixture (No-ap) on the Expression of Angiotensin II Receptor 1 and Oxidative Stress in Angiotensin II-Stimulated H9C2 Cells

Eun Young Hong [1,†], Tae Yang Kim [2,†], Gwan Ui Hong [1], Hanna Kang [2], Jung-Yun Lee [3], Jae Yeo Park [1], Se-Chan Kim [4], Young Ho Kim [5], Myung-Hee Chung [6], Young-In Kwon [2,*] and Jai Youl Ro [1,7,*]

1 Life & Science Research Center, Hyunsung Vital Co. Ltd., Seoul 07255, Korea; h_medical@naver.com (E.Y.H.); gwanuihong@gmail.com (G.U.H.); legolas018@naver.com (J.Y.P.)
2 Department of Food and Nutrition, Hannam University, Daejeon 34054, Korea; xodid5606@naver.com (T.Y.K.); hanna9506@hanmail.net (H.K.)
3 Natural Products Institute, Proteinworks Co. Ltd., Daejeon 07255, Korea; seembeeks@hanmail.net
4 Department of Bio Quality Control, Korea Bio Polytechnic, Chungnam 32943, Korea; sechage@kopo.ac.kr
5 Department of Pharmacy, Choongnam National University, Daejeon 34134, Korea; yhk@cnu.ac.kr
6 Gil Hospital and Lee Gil Ya Cancer & Diabetes Institute, Gachon University, Inchon 21999, Korea; mhchung@gachon.ac.kr
7 Department of Pharmacology, Sungkyunkwan University School of Medicine, Suwon 03063, Korea
* Correspondence: youngk@hnu.kr (Y.-I.K.); jyro426@skku.edu (J.Y.R.); Tel.: +82-42-629-8790 (Y.-I.K.); +82-2-2628-5105 (J.Y.R.); Fax: +82-42-629-8789 (Y.-I.K.); +82-70-4854-4853 (J.Y.R.)
† These authors contributed equally to this work.

check for updates

Received: 21 December 2018; Accepted: 22 January 2019; Published: 23 January 2019

Abstract: Hypertension is a major risk factor for the development of cardiovascular diseases. This study aimed to elucidate whether the natural product mixture No-ap (NA) containing *Pine densiflora*, *Annona muricate*, and *Monordica charantia*, or its single components have inhibitory effects on hypertension-related molecules in Angiotensin II (Ang II)-stimulated H9C2 cells. Individual functional components were isolated and purified from NA using various columns and solvents, and then their structures were analyzed using ESI–MS, ^{1}H-NMR, and ^{13}H-NMR spectra. H9C2 cells were stimulated with 300 nM Ang II for 7 h. NA, telmisartan, ginsenoside, roseoside (Roseo), icariside E4 (IE4), or a combination of two components (Roseo and IE4) were administered to the cells 1 h before Ang II stimulation. The expression and activity of hypertension-related molecules or oxidative molecules were determined using RT-PCR, western blot, and ELISA. Ang II stimulation increased the expression of Ang II receptor 1 (AT1), tumor necrosis factor-α (TNF-α), monocyte chemoattractant protein-1 (MCP-1), tumor growth factor-β (TGF-β) mRNA, and nicotinamide adenine dinucleotide phosphate (NADPH) oxidase activity and the levels of hydrogen peroxide (H_2O_2) and superoxide anion ($\bullet O_2^-$) and reduced anti-oxidant enzyme activity. NA significantly improved the expression or activities of all hypertension-related molecules altered in Ang II-stimulated cells. Roseo or IE4 pretreatment either decreased or increased the expression or activities of all hypertension-related molecules similar to NA, but to a lesser extent. The pretreatment with a combination of Roseo and IE4 (1:1) either decreased or increased the expression of all hypertension-related molecules, compared to each single component, revealing a synergistic action of the two compounds. Thus, the combination of single components could exert promising anti-hypertensive effects similar to NA, which should be examined in future animal and clinical studies.

Keywords: cardiomyocytes (H9C2 cells); angiotensin II; hypertension; No-ap (natural product mixture); reactive oxygen species; roseoside; icariside E4

1. Introduction

Hypertension is a major risk factor for the development of cardiovascular diseases including coronary artery disease, stroke, heart failure, peripheral vascular disease [1]. Angiotensin II (Ang II) is a vasoactive peptide of the renin–angiotensin system (RAS). The cellular effects of Ang II are mediated by at least two receptors, Ang II receptor 1 (AT1) and Ang II receptor 2 (AT2). Ang II, through AT1 or AT2, plays a key role in blood pressure homeostasis [2]. Ang II binds AT1 to induce NADPH oxidase activation and leads to the production of reactive oxygen species (ROS) [3–5]. Thus, the increasing of ROS is involved in cardiovascular diseases-related changes such as hypertrophy, fibrosis, tissue inflammation, or vascular remodeling in the heart [4–6].

Inhibition of the RAS through AT receptor blockers (ARBs) can prevent cardiovascular disease-related events [7]. ARBs (for example, losartan and telmisartan), which are a new class of approved anti-hypertensive agents, prevent the hypertensive effects of Ang II by the selective blockade of AT1 [8,9]. However, ARBs produce undesirable side effects such as headache, fatigue, and dizziness [7]. Thus, we have much interest in the search for natural products with anti-hypertensive effects and reduced side effects.

Our company (Hyunsung Vital Co. Ltd., Seoul, Korea) has manufactured a natural product complex (No-ap, NA) expected to downregulate blood pressure. NA contains three natural materials, i.e., *Pinus densiflora*, *Annona muricata* L., and *Monordica charantia*.

P. densiflora is widely distributed around the world, particularly in Korea and Japan [10]. Pine bark or needle have been reported to be effective scavengers of ROS [11,12] and to have a suppressive effect on the expression of pro-inflammatory mediators [13]. Pine needles have anti-hypertensive effects [14]. *A.a muricata* L., which is known as graviola or guanabana, is widely found in India, South and Central America, tropical West Africa, and Asia [15]. It has been reported that a decoction made from *A. muricata* can be used for hypertension therapy and that the plant extract has anti-oxidative and anti-hypertensive properties [16]. *M. charantia* L. is a common vegetable in Okinawa, where it has been recently used in the therapy of hypertension, diabetes, and dyslipidemia [17], and its phenolic extract has inhibitory properties against angiotensin-1-converting enzyme, hypertension, and oxidative stress [18]. Thus, these findings imply that NA may have anti-hypertensive effects, as it is also supported by its wide use as a folk remedy and by laboratory experiments [14–16].

Although there are reports that the different constituents of NA, i.e., *P. densiflora* needle, *A. muricata*, and *M. charantia* may have anti-hypertensive properties, the single components of this mixture have not been isolated and examined [14–16]. This study investigated for the first time whether NA has inhibitory effects on the hypertension-related molecules in Ang II-stimulated H9C2 cells. After confirmation of the anti-hypertensive effects, this study aimed to identify the single functional components of NA and to investigate whether they have anti-hypertensive properties individually. We observed that the pretreatment with a combination of roseoside and icariside E4, which showed strong activity among the five single components identified in NA, had anti-hypertensive effects by downregulating ROS generated via the expression of AT1 and the activity of NADPH oxidase.

2. Results

2.1. *Effects of NA on the Expression of Hypertension-Related Molecules in Ang II-Stimulated H9C2 Cells*

AT1 is an important effector controlling blood pressure (BP) and blood volume in the cardiovascular system [3]. We first examined the effects of NA on AT1 expression in Ang II-stimulated H9C2 cells. AT1 expression was increased in the Ang II-stimulated H9C2 cells, compared with negative control

(NC, treated with phosphate-buffered saline) cells. NA (60, 100, 200 μg/mL) reduced AT1 expression in a dose-dependent manner (Figure 1A). A high dose of NA reduced AT1 expression similar to telmisartan (Telmis), which is known as an AT1 blocker preventing Ang II-induced oxidative stress and vascular remodeling in hypertension [9]. Pretreatment with 200 μg/mL (corresponding to the high dose of NA) of ginsenoside (Gin), which was used as one of the natural positive controls for the natural product mixture (NA), had no effect on AT1 expression in Ang II-stimulated H9C2 cells.

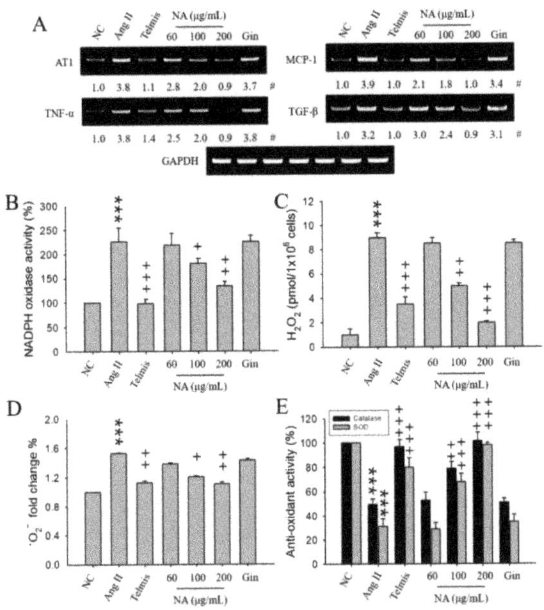

Figure 1. Effects of the natural product mixture (No-ap, NA) on the expression of hypertension-related molecules or oxidative stress in the Ang II-stimulated H9C2 cells. H9C2 cells (1×10^6 cells) were stimulated with 300 nM Ang II for 7 h. No-ap (NA), telmisartan (Telmis), or ginsenoside (Gin) were administered 1 h before Ang II stimulation. The expression of AT1, TNF-α, MCP-1, TGF-β was determined in mRNA extracts isolated from H9C2 cells using RT-PCR. The activity of NADPH oxidase, catalase, and SOD, and the amounts of H_2O_2 and $\bullet O_2^-$ were determined in cell lysates isolated from H9C2 cells using an ELISA kit. The reactions were analyzed using an ELISA plate reader at 450 nm for the activities of NADPH oxidase and SOD and $\bullet O_2^-$ amounts, and at 590 nm for H_2O_2 amounts and catalase activity. (**A**) Expression of AT1 and cytokines. (**B**) Activity of NADPH oxidase. (**C**) Amounts of H_2O_2. (**D**) Amounts (fold change %) of $\bullet O_2^-$. (**E**) Activities of catalase and SOD. #, Numbers below the band images, indicating the mean values (*n* = 4 independent experiments) obtained from the ratio of the band density of each group to those of the corresponding controls and loading control GAPDH. The results represent the mean ± SEM (*n* = 4) obtained from four independent experiments performed in triplicates. NC, negative control; Ang II, angiotensin II stimulation; AT1, angiotensin II receptor 1; TNF-α, tumor necrosis factor-α; MCP-1, monocyte chemoattractant protein-1; TGF-β, tumor growth factor-β; NADPH, nicotinamide adenine dinucleotide phosphate; H_2O_2, hydrogen peroxide; $\bullet O_2^-$, superoxide anion; SOD, superoxide dismutase. ***, $p < 0.001$ versus the NC. $^+$, $p < 0.05$; $^{++}$, $p < 0.01$; $^{+++}$, $p < 0.001$ versus the Ang II stimulation.

It has been reported that inflammation has a crucial role in the pathogenesis of hypertension [19–21]. The inflammatory process, with ROS generation and increase in cytokines' releases, is a hallmark of hypertension [20,21]. Thus, in order to investigate whether NA prevents inflammation in Ang II-stimulated H9C2 cells, the expression of inflammatory cytokines was examined. The expression levels of tumor necrosis factor-α (TNF-α), monocyte chemoattractant protein-1 (MCP-1), and tumor growth factor-β (TGF-β) were increased in Ang II-stimulated H9C2 cells (Figure 1A). NA pretreatment significantly suppressed the expression of these inflammatory cytokines caused by Ang II in a dose-dependent manner. A high dose of NA showed stronger inhibitory responses than Telmis. Gin did not show any effects in all cases. Therefore, hereafter, we will not consider Gin effects.

2.2. Effects of NA on Oxidative Stress in Ang II-Stimulated H9C2 Cells

It has been reported that Ang II-induced hypertension by ROS generated via nicotinamide adenine dinucleotide phosphate (NADPH) oxidase [22,23]. Thus, we examined the effects of NA on NADPH oxidase activity in Ang II-stimulated H9C2 cells. NA pretreatment reduced NADPH oxidase activity in Ang II-stimulated H9C2 cells (Figure 1B).

Next, we examined the effects of NA on the generation of ROS or on anti-oxidant enzyme activity in Ang II-stimulated H9C2 cells. The production of hydrogen peroxide (H_2O_2) or superoxide anion ($\bullet O_2^-$) increased by Ang II stimulation was diminished in NA-pretreated cells in a dose-dependent manner (Figure 1C,D). The activities of catalase or superoxide dismutase (SOD) were decreased in Ang II-stimulated H9C2 cells. NA pretreatment increased anti-oxidant enzyme activity (catalase and SOD) in Ang II-stimulated H9C2 cells in a dose-dependent manner, similar to Telmis (Figure 1E).

2.3. Purification and Identification of Bioactive Ingredients in NA

As a natural product mixture, NA contains *P. densiflora* (75.0%), *A. muricata* (12.5%), *M. charantia* (12.5%). NA was extracted with methanol and successively partitioned with ethyl acetate and n-BuOH to afford ethyl acetate, n-BuOH, and water fractions. The n-BuOH fraction was subjected to silica gel and YMC RP-18 silica gel column chromatography, and five compounds (**1**–**5**) were identified. Their spectroscopic data and comparisons with previous data confirmed that these compounds were roseoside (**1**) [24], isolariciresinol 9-*O*-β-D-xyloside (**2**) [25], massonianoside B (**3**) [26], icariside E4 (**4**) [27], and nicotiflorin (**5**) [28] (Figure 2, Supplementary Materials Figures S1–S15).

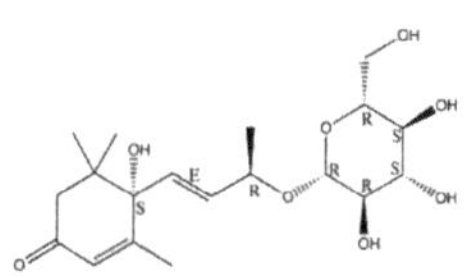

1

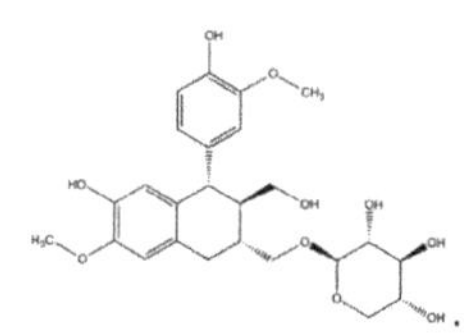

2

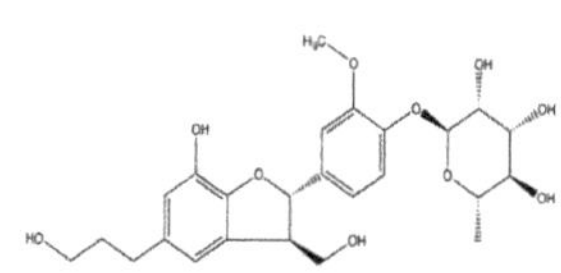

3

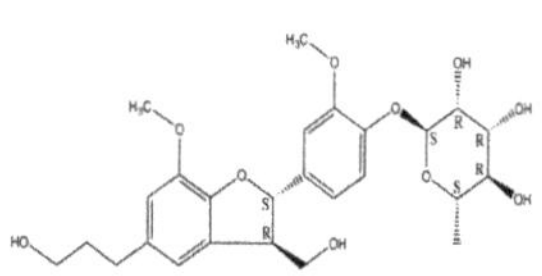

4

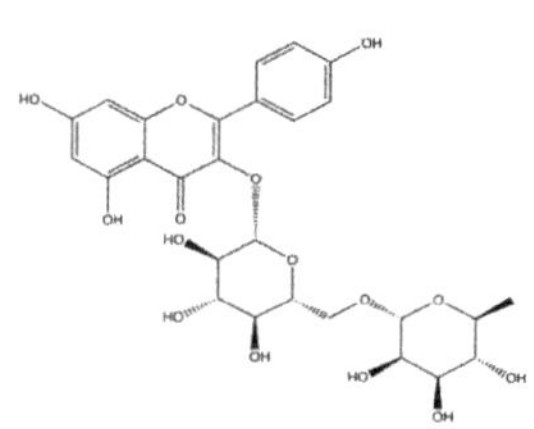

5

Figure 2. Chemical structures of the compounds isolated from NA. NA (1.9 kg) was fractioned using various organic chemicals and columns as described in "Materials and Methods". Finally, five single components were identified from NA, and then the structure of each component was analyzed using ESI–MS, ^{1}H-NMR, and ^{13}C-NMR spectra. The inhibitory effects on the expression of AT1 of each fraction separated with organic chemicals were determined (data not shown). Structures of (**1**), (**2**), (**3**), (**4**), and (**5**) indicate roseoside, isolariciresinol 9-*O*-β-D-xyloside, massonianoside B, icariside E4, and nicotiflorin, respectively. The molecular formulas of the components roseoside and icariside E4, which demonstrated biological activity, re C19H30O8 (MW, 386.1941) and C26H34O10 (MW, 506.5480), respectively.

2.4. *Effects of the Isolated Components Roseoside and Icariside E4 on the Expression of Hypertension-Related Molecules in Ang II-Stimulated H9C2 Cells*

We first confirmed that two single components [roseoside (Roseo) and icariside E4 (IE4)], among the five isolated components, had biological activity. Three components (isolariciresinol 9-*O*-β-D-xyloside, massonianoside B, and nicotiflorin) did not have any effects on the expression of AT1 (data not shown). Roseo or IE4 (20, 30, 50 μg/mL) pretreatment significantly reduced the expression of all hypertension-related molecules (mRNA and protein) in Ang II-stimulated H9C2 cells in a dose-dependent manner, compared to Ang II stimulation alone (Figure 3). Treatment with 70 μg/mL of Roseo or 100 μg/mL of IE4 reduced the expression of all hypertension-related molecules by approximately 48% and 50%, respectively, compared to Ang II stimulation. However, only Roseo showed downregulating activity at a dose above 70 μg/mL (data not shown). Both components showed inhibitory effects on the expression of all hypertension-related molecules, although neither one exhibited inhibitory effects similar to those of Telmis or NA. Thus, we tried a combination pretreatment. A ratio of 1:1 for the combination pretreatment of Roseo and IE4 was used in preliminary experiments. The combination of Roseo and IE4 in a 1:1 ratio (each used in a dose of 25 μg/mL; total dose, 50 μg/mL) reduced the expression of all hypertension-related molecules by approximately 50–72% in Ang II-stimulated H9C2 cells. Thus, the combination pretreatment of Roseo and IE4 showed a strong synergistic inhibitory effect on the expression of all hypertension-related molecules in Ang II-stimulated H9C2 cells.

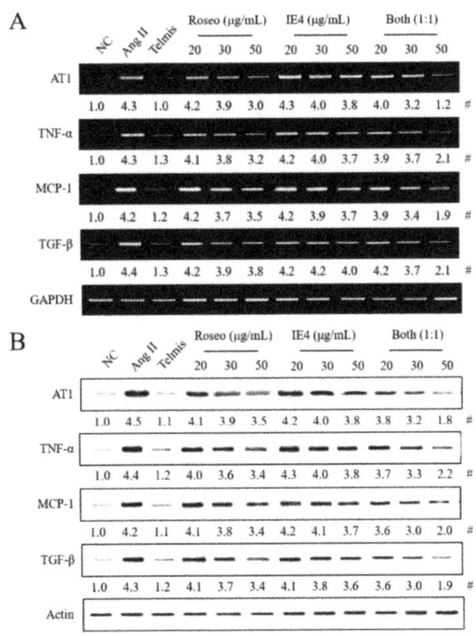

Figure 3. Effects of roseoside and icariside E4, alone or in combination, on the expression of hypertension-related molecules in the Ang II-stimulated H9C2 cells. The experimental details of the stimulation of H9C2 cells and their treatment with different compounds were described in Figure 1 legend. Telmis (10 μM), roseoside (Roseo; 20, 30, or 50 μg/mL), icariside E4 (IE4; 20, 30, or 50 μg/mL), and Roseo/E4 combinations (20, 30, or 50 μg/mL, 1:1 ratio) were administered 1 h before Ang II stimulation. The expression of all hypertension-related molecules was determined in mRNA or protein extracts isolated from H9C2 cells using RT-PCR and Western blot, respectively. (**A**) Expression of all hypertension-related molecules' mRNA. (**B**) Protein expression of all hypertension-related molecules. Both, combination of Roseo and IE4 (1:1 ratio; each component used at a dose of 10, 15, 25 μg/mL; total doses for the different mixtures at a 1:1 ratio were 20, 30, 50 μg/mL). Ang II, angiotensin II stimulation. #, Numbers below the band images indicate the mean values ($n = 4$) obtained as described in Figure 1 legend.

2.5. Respective Effects of Roseoside and Icariside E4 on Oxidative stress in Ang II-Stimulated H9C2 Cells

We examined the effects of Roseo or IE4 on NADPH oxidase activity in Ang II-stimulated H9C2 cells. Roseo or IE4 pretreatment reduced NADPH oxidase activity in Ang II-stimulated H9C2 cells (Figure 4A).

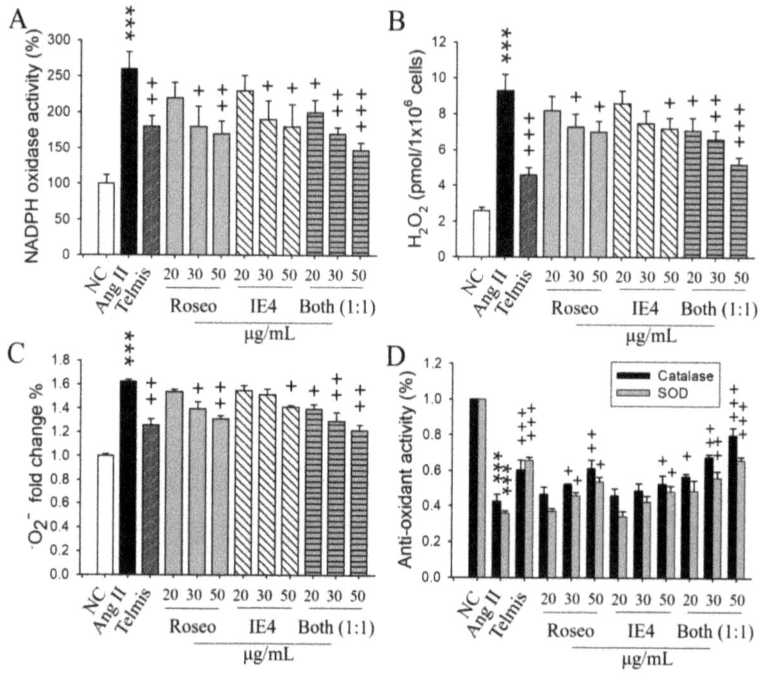

Figure 4. Effects of roseoside, icariside E4, and their combination on oxidative stress in Ang II-stimulated H9C2 cells. The experimental details of the stimulation of H9C2 cells and their treatment with different compounds were described in Figure 1 legend. Telmis (10 μM), Roseo (20, 30, or 50 μg/mL), IE4 (20, 30, or 50 μg/mL), and combinations of Roseo and IE4 (20, 30, or 50 μg/mL, 1:1 ratio) were administered to the cells 1 h before Ang II stimulation. The activities of NADPH oxidase, catalase, and SOD, and the amounts of H_2O_2 and $\bullet O_2^-$ were determined in cell lysates isolated from H9C2 cells using an ELISA kit. The reactions were analyzed using an ELISA plate reader as described in Figure 1 legend. (**A**) Activity of NADPH oxidase. (**B**) Amounts of H_2O_2. (**C**) Amounts (fold change %) of $\bullet O_2^-$. (**D**) Activities of catalase and SOD. The results represent the mean ± SEM ($n = 4$) obtained from four independent experiments performed in triplicates. ***, $p < 0.001$ versus NC. $^+$, $p < 0.05$; $^{++}$, $p < 0.01$; $^{+++}$, $p < 0.001$ versus Ang II stimulation.

Next, we examined the effects of Roseo and IE4 on the generation of ROS and on anti-oxidant enzyme activity in Ang II-stimulated H9C2 cells. The generation of H_2O_2 or $\bullet O_2^-$ increased by Ang II stimulation was diminished after Roseo or IE4 pretreatment in a dose-dependent manner (Figure 4B,C). Both Roseo and IE4 pretreatments increased the anti-oxidant activity (catalase and SOD), which was decreased by Ang II-stimulated H9C2 cells (Figure 4D). Roseo pretreatment showed reduction of ROS and increase of anti-oxidant enzyme activity similar to Telmis, whereas IE4 pretreatment showed smaller effects than Roseo or Telmis.

The combination pretreatment of Roseo and IE4 at a ratio of 1:1 (each used at the dose of 25 μg/mL; total dose, 50 μg/mL) reduced the activity of NADPH oxidase and the generation of H_2O_2 and $\bullet O_2^-$ in a dose-dependent manner in Ang II-stimulated H9C2 cells. The combination pretreatment increased the activity of catalase or SOD in a dose-dependent manner. Thus, the combination pretreatment of Roseo and IE4 at a 1:1 ratio showed stronger synergistic inhibitory effects on the suppression of oxidative stress and on the increase of anti-oxidant enzyme activities in Ang II-stimulated H9C2 cells, compared to the single components (Figure 4).

3. Discussion

We demonstrated that the natural product complex NA suppresses the expression of AT1, the generation of ROS (H_2O_2 and $\bullet O_2^-$), and the expression of inflammatory cytokines (TNF-α, MCP-1, and TGF-β) produced via NADPH oxidase, which are related to hypertension, and that it increases the expression of anti-oxidant enzymes (catalase and SOD) in Ang II-stimulated H9C2 cells. Our data also demonstrate that two components, Roseo and IE4, among five identified components isolated and purified from NA, whose structures were also analyzed in this study, have functional activities in hypertension.

The component Roseo isolated from various plants, including *A. muricata* contained in NA, has a variety of functional activities. It relaxes precontracted aortic rings in an endothelium-dependent manner [29], increases insulin secretion [30], inhibits rat liver microsomal glucose-β-phosphate [31], potentiates the inhibitory activity against angiotensin-converting enzyme, although itself shows no activity against an its enzyme [32], shows inhibitory effects on lipopolysaccharide (LPS)-induced nitric oxide (NO) production in RAW264.7 cells [33], prevents oxidative stress [34], and has a depigmentation effect in melanocytes by inhibiting melanin synthesis [35]. However, it has not been reported yet that Roseo directly suppresses hypertensive effects caused by the upregulation of ROS.

The other component, IE4, isolated and purified from plants, such as *Ulmus pumila* L. and *Tabebuia roseo-alba*, has various functional activities. It inhibits NO production [36], shows anti-oxidant activity [37], demonstrates anti-nociceptive activity in a chemical pain-induced model [38], and may be beneficial in the traditional treatment of Alzheimer's disease by preventing blood–brain barrier damage and inflammatory cell infiltration into the brain [39]. IE4, purified from pine trees, such as *P. densiflora*, *Pinus thunbergii*, and *Pinus morrisonicola* Hayata, also delays the coagulation time by inhibiting thrombin activity [40] and inhibits the release of β-hexosaminidase in RBL-2H3 cells [41]. However, it has not been reported yet that IE4 purified from *P. densiflora*, which is the most abundant component of NA, directly suppresses the hypertensive effects caused by the upregulation of ROS. There is only a report indicating that IE4 isolated from *P. morrisonicola* H. could be a promising anti-hypertensive candidate by blocking voltage-operating Ca^{2+} channels [42].

Hypertension is the most common cardiovascular risk factor. Ang II, which is known as one of many factors causing cardiovascular injury in hypertension, elicits many pathophysiological actions by inducing ROS generation via the activation of vascular NADPH oxidase [20,22,23,43]. Ang II stimulation can increase blood pressure in association with immune response activation and inflammation [21]. Infiltration of inflammatory cells into areas around blood vessels that occurs simultaneously with other events of the inflammatory process, such as the increase of ROS generation and of the levels of cytokines and chemokines, is a hallmark of hypertension [9,21,22]. Thus, our data suggest that NA may suppress blood pressure and vascular remodeling in hypertension mainly caused by RAS through downregulating of ROS produced via AT1 expression and NADPH oxidase activity, as demonstrated by the data showing that NA reduced hypertensive responses through inhibiting downstream pathways via AT1 expression and NADPH oxidase (which generates ROS), and then the generated ROS increased the release of the cytokine TNF-α and the chemokines MCP-1, which is known to induce inflammation and cell infiltration [44], respectively, and increased TGF-β, which is known to induce vascular remodeling in hypertension [9,22]. In addition, our data indicate that NA increased endogenous anti-oxidant enzymes (catalase and SOD) in Ang II-stimulated H9C2 cells. The endogenous anti-oxidant glutathione peroxidase (GPx), which scavenges H_2O_2, is not affected by Ang II stimulation in cardiac fibroblasts [45] and vascular adventitial fibroblasts [46]. Thus, we did not check GPx activity because the catalase also scavenges H_2O_2. However, it is necessary to investigate the differences between cell types.

Our observations also suggest that the components Roseo and IE4, isolated and purified from NA, may independently suppress hypertension-related molecules such as ROS, cytokines and chemokines, and TGF-β, which are typically associated with oxidative stress, inflammation, and vascular remodeling in hypertension, by increasing endogenous anti-oxidant enzymes in Ang II-stimulated H9C2 cells. In addition, we also show that combinations of Roseo and IE4 (1:1 ratio) may have strong synergistic or additive effects.

In conclusion, these in vitro results support the conclusion that the functional compounds Roseo and IE4 have significant anti-hypertensive effects on Ang II-stimulated H9C2 cells. Our data suggest that Roseo and IE4 from NA, which have anti-oxidant, anti-inflammatory, and anti-vascular remodeling properties in hypertension and less side effects than the whole mixure NA, have a potential as health functional food supplements for hypertension and should be further evaluated in animal and clinical models.

4. Materials and Methods

4.1. Materials

NA, a natural product mixture, was donated by Hyunsung Vital Co. Ltd. (Seoul, Korea) which makes a variety of healthy functional foods. NA contains *P. densiflora* (75.0%), *A. muricata* (12.5%), and *M. charantia* (12.5%). These plants were extracted using hot water (90 °C) for 24 h, and the extract was evaporated by a Liquefied extractor (Hyunsung Vital Co. Ltd., Seoul, Korea) to yield a powder of NA. This powder was named NA (No-ap), natural product mixture.

4.2. Purification and Identification of Bioactive Ingredients

NA (1.9 kg) was extracted using methanol (4 L, 95%) at room temperature for 2 days. The methanol extract (556.4 g) was concentrated under pressure, dissolved in distilled water (2 L), and successively partitioned with ethyl acetate and n-BuOH to afford ethyl acetate (23.0 g, A), n-BuOH (122.1 g, B), and water fractions.

The n-BuOH extract (120 g) was separated by vacuum liquid chromatography using a silica gel column with a gradient solvent mixture of $CHCl_3$–MeOH (30:1, 25:1, 20:1, 15:1, 10:1, 8:1, 6:1, 4:1, 2:1, 1:1, and 100% MeOH) to afford 11 subfractions (B-1 to B-11). Next, subfraction B-6 (2.1 g) was subjected to YMC RP-C18 silica gel column and was eluted with a solvent mixture of MeOH–H_2O (1:2), yielding compound 1 (20.0 mg), compound 2 (10.0 mg), compound 3 (30.0 mg), and compound 4 (10.0 mg). Further purification of subfraction B-9 via YMC RP-C18 silica gel column, using mixtures of MeOH−H_2O (1:1.5), and preparative HPLC yielded compound 5 (10.0 mg).

4.3. Cardiomyocytic H9C2 Cell Line Culture

The rat cardiomyocytic H9C2 cell line (H9C2 cells) was obtained from the Korean Cell line Bank (KCLB, Seoul, Korea). H9C2 cells were grown in Dulbecco's modified eagle medium (DMEM) (HyClone, Logan, UT, USA) supplemented with 1% L-glutamine, 1% antibiotic penicillin/streptomycin solution (Sigma-Aldrich, St. Louis, MO, USA), and 10% fetal bovine serum (HyClone, Logan, UT, USA). H9C2 cells were maintained at 37 °C in a humidified atmosphere with CO_2, and the media were replaced every 3 days [47].

4.4. Cell line Stimulation and Treatment

H9C2 cells (1×10^6 cells) were stimulated with 300 nM angiotensin II (Ang II; Sigma-Aldrich, St. Louis, USA) and then incubated for 7 h [47]. The cells were centrifuged (470× *g*, 3 min) to separate the supernatants and pellet the cells. The cells were used to determine the expression of all molecules related to hypertension. NA (60, 100 or 200 μg/mL), telmisartan (Telmis; 10 μM), ginsenoside (Gin; 200 μg/mL), roseoside (Roseo; 20, 30 or 50 μg/mL), icariside E4 (IE4; 20, 30 or 50 μg/mL), the last two as purified components isolated from NA, and combinations of Roseo and IE4 (1:1 ratio,

20, 30, or 50 μg/mL) were administered to the cells 1 h before Ang II stimulation. As a negative control (NC), phosphate-buffered saline (PBS) was used, which was also Ang II solvent. Telmis, which is a drug used in the clinic, was used as a positive control. The optimal concentrations of Ang II stimulation, NA, Gin, Roseo, and IE4 were determined in preliminary experiments (data not shown). Gin was used as a natural positive control for the natural product mixture (NA).

4.5. Reverse Transcription-Polymerase Chain Reaction (RT-PCR)

Total mRNA was isolated from H9C2 cells (1×10^6 cells) using TRIzol reagent (Invitrogen, Life Technologies Ltd., Waltham, CA, USA). RT-PCR was performed in a final volume of 20 μL, using a high-capacity cDNA Reverse Transcription kit (Applied Biosystems, Foster City, CA, USA) and a G-taq kit (Cosmogenetech, Seoul, Korea) in an automated thermal cycler (Bio-Rad, Laboratories, Hercules, CA, USA). The PCR assays were performed for 35 cycles. Each cycle consisted of the following steps: denaturation at 94 °C for 30 s, annealing at 51 °C for 45 s, and extension at 72 °C for 1 min. The results were expressed as a ratio to GAPDH mRNA. The PCR products were analyzed using 1% agarose gel and visualized under UV light after staining with StaySafe nucleic acid gel stain (Real Biotech Corporation, Banqiao, Taiwan) [48].

The primer sequences used were as follows: AT1 sense, 5′-CAT AGG ACT GGG CCT AAC CA-3′; AT1 anti-sense, 5′-GCC GTA AAA CAG AGG GTT CA-3′; TNF-α sense, 5′-TTC TGT CCC TTT CAC TCA CTG G-3′; TNF-α anti-sense, 5′-TTG GTG GTT TGC TAC GAC GTG G-3′; MCP-1 sense, 5′-GAA GGA ATG GGT CCA GAC AT-3′; MCP-1 anti-sense, 5′-ACG GGT CAA CTT CAC ATT CA-3′; TGF-β sense, 5′-CTC TCC ACC TGC AAG ACC AT-3′; TFG-β anti-sense, 5′-CTG CCG TAC AAT TCC AGT GA-3′; GAPDH sense, 5′-AAC TTT GGC ATT GTG GAA GG-3′; GAPDH anti-sense, 5′-ACA CAT TGG GGG TAG GAA CA-3′.

4.6. Determination of the Activity of NADPH Oxidase, Superoxide Dismutase, and Catalase and the Levels of Hydrogen Peroxide and Superoxide Anion

The activities of NADPH oxidase, catalase, and SOD, and the amounts of H_2O_2 and $\bullet O_2^-$ were measured in the lysates of H9C2 cells stimulated with Ang II using a NADPH oxidase assay kit, a H_2O_2 assay kit (Abcam, Cambridge, UK), a $\bullet O_2^-$ assay kit, a SOD assay kit (Cell Biolabs. Inc., San Diego, CA, USA), and a catalase activity kit (Biovision, Milpitas, CA, USA), respectively. Briefly, H9C2 cells (1×10^6 cells) were washed three times with PBS and placed in lysis buffer (PBS containing 1% Triton X-100). Standard diluents (100 μL), lysates obtained from each sample (100 μL), and 100 μL of reaction mixture (50 μL enzyme working solution and 50 μL probe) were added in 96-well plates; the plates were then incubated on a plate shaker at room temperature for the time periods indicated (30 min for NADPH oxidase, SOD, and catalase; 1 h for H_2O_2 and $\bullet O_2^-$). The optical density was read at 450 nm for NADPH oxidase, $\bullet O_2^-$, and SOD, and at 590 nm for H_2O_2 and catalase. Standard curves were made using serial dilutions of a standard sample, and then the activity was calculated according to the manufacturer's instructions. The lowest detection limit for NADPH oxidase was below 20 pg/mL, for H_2O_2 it was 0.04 pmol/μL, for $\bullet O_2^-$ and SOD it was 1.2 mU/μL, and for catalase it was 0.078 pg/mL.

4.7. Western Blot Analysis

H9C2 cells (1×10^6 cells) harvested from Ang II-stimulated cells were suspended in a low-salt lysis buffer [50 mM Tri-HCl (pH 7.9), 1.0 mM EDTA, 150 mM NaCl, 1.0% NP40, 5 mM NaF, 0.25% Na deoxycholate, 2 mM $NaVO_3$, protease inhibitors cocktail] and allowed to swell on ice for 30 min. The cells were then homogenized using a micropipette. After centrifugation, the supernatants obtained from the cell extracts were analyzed by 10% SDS-polyacrylamide gel electrophoresis and electrophoretically transferred to nitrocellulose membranes (Amersham Biosciences, Piscataway, NJ, USA). The membranes were washed with PBS containing 0.1% Tween 20 (PBST) and then blocked for 1 h in PBST containing 5% skim milk. After washing the membranes with PBST, they were treated with primary antibodies against actin, AT1, TNF-α (Cell Signaling

Technology, Beverly, MA, USA), MCP-1, or TGF-β (Abcam, Cambridge, UK) diluted with PBST (1:1000). The membranes were washed with PBST and treated with horseradish peroxidase (HRP)-conjugated goat anti-mouse or HRP-conjugated goat anti-rabbit IgG (diluted to 1:5000) (Bethyl Laboratories, Montgomery, TX, USA) in PBST for 1 h. After washing, the protein bands were visualized by Enhanced Chemi-Luminescence (ECL; Amersham Biosciences, Piscataway, NJ, USA) using a chemiluminometer (Bio-Rad, Laboratories, CA, USA) [48].

4.8. Statistical Analysis

The experimental data are shown as means ± SEM ($n = 4$). The unpaired Student's t-test was used to compare two groups. Multiple-group comparisons were performed using two-way ANOVA followed by Scheffe's post-hoc test, using the SPSS software (SPSS Inc., Chicago, IL, USA). Values of $p < 0.05$ were considered to indicate statistical significance. Densitometry analyses of Western blots and RT-PCR were performed with Quantity One (version 4.6.3; Bio-Rad, Hercules, CA, USA) and the results are indicated as means ± SEM ($n = 4$), obtained from the ratio of each band density to those of the control and the loading control of four independent experiments.

Supplementary Materials: The following are available online at http://www.mdpi.com/1420-3049/24/3/414/s1, Figure S1: ^{1}H-NMR spectra of roseoside (**1**), Figure S2: ^{13}C-NMR spectra of roseoside (**1**), Figure S3: ESI–MS spectra of roseoside (**1**), Figure S4: ^{1}H-NMR spectra of (−)-marmesinin (**2**). Figure S5: ^{13}C-NMR spectra of isolariciresinol 9-*O*-β-D-xyloside (**2**), Figure S6: ESI-MS spectra of isolariciresinol 9-*O*-β-D-xyloside (**2**), Figure S7: ^{1}H-NMR spectra of massonianoside B (**3**), Figure S8: ^{13}C-NMR spectra of massonianoside B (**3**), Figure S9: ESI-MS spectra of massonianoside B (**3**), Figure S10: ^{1}H-NMR spectra of icariside E4 (**4**), Figure S11: ^{13}C-NMR spectra of icariside E4 (**4**), Figure S12: ESI-MS spectra of icariside E4 (**4**), Figure S13: ^{1}H-NMR spectra of nicotiflorin (**5**), Figure S14: ^{13}C-NMR spectra of nicotiflorin (**5**), Figure S15: ESI-MS spectra of nicotiflorin (**5**).

Author Contributions: J.Y.R. designed the study and directed its implementation, including the study's analytic strategy, and prepared the Discussion section of the text. Y.-I.K. helped conduct the literature review and prepare the Methods and the Discussion sections of the text. Y.H.K., S.-C.K., J.-Y.L., H.K., and T.Y.K. prepared the products to be tested and participated in the preparation of the manuscript. M.-H.C. helped conduct the literature review and designed the study. E.Y.H., G.U.H., and J.Y.P. conducted the cell experiments and analyzed the data. All the authors read and approved the final manuscript.

Funding: This study was supported by a research grant from Life & Science Research Center, Hyunsung Vital Co. Ltd., Seoul, Korea 07255, and from Hannam University in 2018, Daejeon 34054, Korea.

Conflicts of Interest: The authors declare no conflict of interest.

References

1. Lackland, D.T.; Weber, M.A. Global burden of cardiovascular disease and stroke: Hypertension at the core. *Can. J. Cardiol.* **2015**, *31*, 569–571. [CrossRef] [PubMed]
2. Nguyen Dinh Cat, A.; Montezano, A.C.; Burger, D.; Touyz, R.M. Angiotensin II, NADPH oxidase, and redox signaling in the vasculature. *Antioxid. Redox Signal.* **2013**, *19*, 1110–1120. [CrossRef] [PubMed]
3. Zablocki, D.; Sadoshima, J. Angiotensin II and oxidative stress in the failing heart. *Antioxid. Redox Signal.* **2013**, *19*, 1095–1109. [CrossRef] [PubMed]
4. Nakagami, H.; Takemoto, M.; Liao, J.K. NADPH oxidase-derived superoxide anion mediates angiotensin II-induced cardiac hypertrophy. *J. Mol. Cell. Cardiol.* **2003**, *35*, 851–859. [CrossRef]
5. Takimoto, E.; Kass, D.A. Role of oxidative stress in cardiac hypertrophy and remodeling. *Hypertension* **2007**, *49*, 241–248. [CrossRef] [PubMed]
6. Dong, B.; Liu, C.; Xue, R.; Wang, Y.; Sun, Y.; Liang, Z.; Fan, W.; Jiang, J.; Zhao, J.; Su, Q.; et al. Fisetin inhibits cardiac hypertrophy by suppressing oxidative stress. *J. Nutr. Biochem.* **2018**, *62*, 221–229. [CrossRef] [PubMed]
7. Oparil, S.; Williams, D.; Chrysant, S.G.; Marbury, T.C.; Neutel, J. Comparative efficacy of olmesartan, losartan, valsartan, and irbesartan in the control of essential hypertension. *J. Clin. Hypertens.* **2001**, *3*, 283–291. [CrossRef]
8. Burnier, M.; Brunner, H.R. Angiotensin II receptor antagonists. *Lancet* **2000**, *355*, 637–645. [CrossRef]

9. Shang, P.; Liu, T.; Liu, W.; Li, Y.; Dou, F.; Zhang, Y.; Sun, L.; Zhang, T.; Zhu, Z.; Mu, F.; et al. Telmisartan improves vascular remodeling through ameliorating prooxidant and profibrotic mechanisms in hypertension via the involvement of transforming growth factor-β1. *Mol. Med. Reports.* **2017**, *16*, 4537–4544. [CrossRef] [PubMed]
10. Wang, D.; Chen, J.; Xu, Z.; Qiao, X.; Huang, L. Disappearance of polycyclic aromatic hydrocarbons sorbed on surfaces of pine [*Pinua thunbergii*] needles under irradiation of sunlight: Volatilization and photolysis. *Atmos. Environ.* **2005**, *39*, 4583–4591. [CrossRef]
11. Yu, L.; Zhao, M.; Wang, J.S.; Cui, C.; Yang, B.; Jiang, Y.; Zhao, Q. Antioxidant, immunomodulatory and anti-breast cancer activities of phenolic extract from pine (*Pinus massoniana* Lamb) bark. *Innov. Food Sic. Emerg. Technol.* **2009**, *9*, 122–128. [CrossRef]
12. Kim, J.-S. Evaluation of in vitro antioxidant activity of the water extract obtained from dried pine needle (*Pinus densiflora*). *Prev. Nutr. Food Sci.* **2018**, *23*, 134–143. [CrossRef]
13. Venkatesan, T.; Choi, Y.-W.; Lee, J.; Kim, Y.-K. *Pinus densiflora* needle supercritical fluid extract suppresses the expression of pro-inflammatory mediators iNOS, IL-6 and IL-1β, and activation of inflammatory STAT1 and STAT3 signaling proteins in bacterial lipopolysaccharide-challenged murine macrophages. *Daru. J. Pharmaceut. Sci.* **2017**, *25*, 18–27.
14. Cheong, H.S.; Lim, D.Y. Pine needle extracts inhibit contractile responses of the isolated rat aortic strips. *Nat. Prod. Sci.* **2010**, *16*, 123–132.
15. Moghadamtousi, S.Z.; Fadaeinasab, M.; Nikzad, S.; Mohan, G.; Ali, H.M.; Kadir, H.A. *Annona muricata* (Annonaceae): A review of its traditional uses, isolated acetogenin and biological activities. *Int. J. Mol. Sci.* **2015**, *16*, 15625–15658. [CrossRef]
16. Adefegha, S.A.; Oyeleye, S.I.; Oboh, G. Distribution of phenolic contents, antidiabetic potentials, antihypertensive properties, and antioxidative effects of Soursop (*Annona muricata* L.) fruit parts in vitro. *Biochem. Res. Int.* **2015**. [CrossRef]
17. Ojewole, J.A.; Adewole, S.O.; Olayiwola, G. Hypoglycaemic and hypotensive effects of *Momordica charantia* Linn (*Cucurbitaceae*) whole-plant aqueous extract in rats. *Cardiovasc. J. S. Afr.* **2006**, *17*, 227–232.
18. Shodehinde, S.A.; Adefegha, S.A.; Oboh, G.; Oyeleye, S.I.; Olasehinde, T.A.; Nwanna, E.E.; Adedayo, B.C.; Boligon, A.A. Phenolic composition and evaluation of methanol and aqueous extracts of bitter gourd (*Momordica charantia* L.) leaves on angiotensin-I-converting enzyme and some pro-oxidant-induced lipid peroxidation in vitro. *J. Evid. Based Complementary Altern. Med.* **2016**, *21*, 67–76. [CrossRef]
19. Savoia, C.; Schiffrin, E.L. Inflammation in hypertension. *Curr. Opin. Nephrol. Hypertens.* **2006**, *15*, 152–158. [CrossRef] [PubMed]
20. Dinh, Q.N.; Drummond, G.R.; Sobey, C.G.; Chrissobolis, S. Roles of inflammation, oxidative stress, and vascular dysfunction in hypertension. *Biomed. Res. Int.* **2014**. [CrossRef]
21. Agita, A.; Alsagaff, M.T. Inflammation, Immunity, and Hypertension. *Acta. Med. Indones.* **2017**, *49*, 158–165. [PubMed]
22. Prathapan, A.; Vineetha, V.P.; Abhilash, P.A.; Raghu, K.G. *Boerhaavia diffusa* L. attenuates angiotensin II-induced hypertrophy in H9C2 cardiac myoblast cells via modulating oxidative stress and down-regulating NF-κβ and transforming growth factor β1. *Br. J. Nutr.* **2013**, *110*, 1201–1210. [CrossRef]
23. Gray, S.P.; Jandeleit-Dahm, K.A. The role of NADPH oxidase in vascular disease–hypertension, atherosclerosis & stroke. *Curr. Pharm. Des.* **2015**, *21*, 5933–5944. [PubMed]
24. Kuang, H.-X.; Yang, B.-Y.; Xia, Y.-G.; Feng, W.-S. Chemical constituents from the flower of *Datura metel* L. *Arch. Pharm. Res.* **2008**, *31*, 1094–1097. [CrossRef]
25. Feng, W.-S.; Bi, Y.-F.; Zheng, X.-K.; Wang, S.-L.; Li, J. Studies on the lignin chemical constituents from pine needles of *Pinus massomiana* Lamb. *Acta Pharmaceutica Sinica.* **2003**, *38*, 199–202. [PubMed]
26. Nakanishi, T.; Iida, N.; Inatomi, Y.; Murata, H.; Inada, A.; Murata, J.; Lang, F.A.; Iinuma, M.; Tanaka, M. Neolignan and flavonoid glycosides in *Juniperus communis* var. depressa. *Phytochemistry* **2004**, *65*, 207–213. [CrossRef] [PubMed]
27. Kwon, J.H.; Kim, J.H.; Choi, S.E.; Park, K.H.; Lee, M.W. Inhibitory effects of phenolic compounds from needles of *Pinus densiflora* on nitric oxide and PGE2 production. *Arch. Pharm. Res.* **2010**, *33*, 2011–2016. [CrossRef]
28. Hou, W.-C.; Lin, R.-D.; Lee, T.-H.; Huang, Y.-H.; Hsu, F.-L.; Lee, M.-H. The phenolic constituents and free radical scavenging activities of *Gynura formosana* Kiamnra. *J. Sci. Food Agri.* **2005**, *85*, 615–621. [CrossRef]

29. Lee, T.H.; Wang, G.J.; Lee, C.K.; Kuo, Y.H.; Chou, C.H. Inhibitory effects of glycosides from the leaves of *Melaleuca quinquenervia* on vascular contraction of rats. *Planta Med.* **2002**, *68*, 492–496. [CrossRef] [PubMed]
30. Frankish, N.; de Sousa Menezes, F.; Mills, C.; Sheridan, H. Enhancement of insulin release from the beta-cell line INS-1 by an ethanolic extract of *Bauhinia variegata* and its major constituent roseoside. *Planta Med.* **2010**, *76*, 995–997. [CrossRef]
31. Bermudez, J.; Rodriguez, M.; Hasegawa, M.; Gonzalez-Mujica, F.; Duque, S.; Ito, Y. (6R,9S)-6″-(4″-hydroxybenzoyl)-roseoside, a new megastigmane derivative from *Ouratea polyantha* and its effect on hepatic glucose-6-phosphatase. *Nat. Prod. Commun.* **2012**, *7*, 973–976. [PubMed]
32. Simaratanamongkol, A.; Umehara, K.; Noguchi, H.; Panichayupakaranant, P. Identification of a new angiotensin-converting enzyme (ACE) inhibitor from Thai edible plants. *Food Chem.* **2014**, *165*, 92–97. [CrossRef] [PubMed]
33. Huong, N.T.; Vien, L.T.; Hanh, T.T.; Dang, N.H.; Thanh, N.V.; Cuong, N.X.; Nam, N.H.; Truong, L.H.; Ban, N.K.; Kiem, P.V.; et al. Triterpene saponins and megastigmane glucosides from *Camellia bugiamapensis*. *Bioorg. Med. Chem. Lett.* **2017**, *27*, 557–561. [CrossRef] [PubMed]
34. Pinto, J.; Spinola, V.; Llorent-Martinerz, E.J.; Fernandez-de Cordova, M.L.; Molina-Garcia, L.; Castilho, P.C. Polyphenolic profile and antioxidant activities of Madeiran elderberry (*Sambucus lanceolata*) as affected by simulated in vitro digestion. *Food Res. Int.* **2017**, *100*, 404–410. [CrossRef]
35. Lee, Y.R.; Park, J.H.; Castaneda Molina, R.; Nam, Y.H.; Lee, Y.G.; Hong, B.N.; Baek, N.I.; Kang, T.H. Skin depigmenting action of silkworm (*Bombyx mori* L.) droppings in zebrafish. *Arch. Dermatol. Res.* **2018**, *310*, 245–253. [CrossRef] [PubMed]
36. Joo, T.; Sowndhararajan, K.; Hong, S.; Lee, J.; Park, S.Y.; Kim, S.; Jhoo, J.W. Inhibition of nitric oxide production in LPS-stimulated RAW 264.7 cells by stem bark of *Ulmus pumila* L. *Saudi J. Biol. Sci.* **2014**, *21*, 427–435. [CrossRef] [PubMed]
37. Mina, S.A.; Melek, F.R.; Adeeb, R.M.; Hagag, E.G. LC/ESI-MS/MS profiling of *Ulmus parvifolia* extracts and evaluation of its anti-inflammatory, cytotoxic, and antioxidant activities. *Z. Naturforsch. C.* **2016**, *71*, 415–421. [CrossRef]
38. Ferreira-Junior, J.C.; Conserva, L.M.; Lyra Lemos, R.P.; de Omena-Neta, G.C.; Cavalcante-Neto, A.; Barreto, E. Isolation of a dihydrobenzofuran lignan, icariside E4, with an antinociceptive effect from Tabebuia roseo-alba (*Ridley*) Sandwith (*Bignoniaceae*) bark. *Arch. Pharm. Res.* **2015**, *38*, 950–956. [CrossRef] [PubMed]
39. Wang, X.D.; Dubois, R.; Young, C.; Lien, E.J.; Adams, J.D. Heteromeles arbutifolia, a traditional treatment for alzheimer's disease, phytochemistry and safety. *Medicines* **2016**, *3*, 7. [CrossRef] [PubMed]
40. Seo, M.J.; Kang, B.W.; Jeong, Y.K. Identification of a neolignan glycoside from the pine tree, *Pinus densiflora* showed antithrombotic activity. *J. Life Sci.* **2014**, *24*, 873–879. [CrossRef]
41. Hong, S.S.; Jeong, W.; Kim, J.K.; Kwon, J.G.; Lee, J.Y.; Ahn, E.K.; Oh, J.; Seo, D.W.; Oh, J.S. Neolignan inhibitors of antigen-induced degranulation in RBL-2H3 cells from the needles of *Pinus thunbergii*. *Fitoterapia* **2014**, *99*, 347–351. [CrossRef]
42. Chen, G.-H.; Li, Y.-C.; Lin, N.-H.; Kuo, P.-C.; Tzen, J.T.C. Characterization of vasorelaxant principles from the needles of *Pinus morrisonicola* Hayata. *Molecules* **2018**, *23*, 86. [CrossRef] [PubMed]
43. Montezano, A.C.; Nguyen Dinh Cat, A.; Rios, F.J.; Touyz, R.M. Angiotensin II and vascular injury. *Curr. Hypertens. Rep.* **2014**, *16*, 431. [CrossRef]
44. Lee, H.; Kim, K.C.; Hong, Y.M. Changes of bax, Bcl2, chemokine receptor-2, monocyte chemoattractant protein-1, and transforming growth factor β1 genes in the left ventricle of spontaneously hypertensive rat after losartan treatment. *Kor. J. Pediat.* **2018**. [CrossRef] [PubMed]
45. Lijnen, P.J.; van Pelt, J.F.; Fagard, R.H. Downregulation of manganese superoxide dismutase by angiotensin II in cardiac fibroblasts of rats: Association with oxidative stress in myocardium. *Am. J. Hypertens.* **2010**, *23*, 1128–1135. [CrossRef]
46. Lubos, E.; Loscalzo, J.; Handy, D.E. Glutathione peroxidase-1 in health and disease: From molecular mechanisms to therapeutic opportunities. *Antioxid. Redox Signal.* **2011**, *15*, 1957–1997. [CrossRef] [PubMed]
47. Yan, L.; Zhang, J.D.; Wang, B.; Lv, Y.J.; Jiang, H.; Liu, G.L.; Qiao, Y.; Ren, M.; Guo, X.F. Quercetin inhibits left ventricular hypertrophy in spontaneously hypertensive rats and inhibits angiotensin II-induced H9C2 cells hypertrophy by enhancing PPAR-gamma expression and suppressing AP-1 activity. *PLoS ONE* **2013**, *8*. [CrossRef] [PubMed]

48. Hong, G.U.; Kim, N.G.; Kim, T.J.; Ro, J.Y. CD1d expressed in mast cell surface enhances IgE production in B cells by up-regulating CD40L expression and mediator release in allergic asthma in mice. *Cell. Signal.* **2014**, *26*, 1105–1117. [CrossRef]

Sample Availability: Samples of the compounds are available from the corresponding authors (Y.-I.K.; J.Y.R.).

molecules

MDPI

Article

Triterpene-Based Carboxamides Act as Good Inhibitors of Butyrylcholinesterase

Anne Loesche [1], Michael Kahnt [1], Immo Serbian [1], Wolfgang Brandt [2] and René Csuk [1,*]

1 Martin-Luther-University Halle-Wittenberg, Organic Chemistry, Kurt-Mothes-Str. 2, D-06120 Halle (Saale), Germany; anne.loesche@chemie.uni-halle.de (A.L.); michael.kahnt@chemie.uni-halle.de (M.K.); immo.serbian@chemie.uni-halle.de (I.S.)

2 Leibniz Institute of Plant Biochemistry, Bioorganic Chemistry, Weinberg 3, D-06120 Halle (Saale), Germany; wbrandt@ipb-halle.de

* Correspondence: rene.csuk@chemie.uni-halle.de; Tel.: +49-345-55-25660

Academic Editors: Eva E. Rufino-Palomares and José Antonio Lupiáñez
Received: 12 February 2019; Accepted: 2 March 2019; Published: 7 March 2019

Abstract: A set of overall 40 carboxamides was prepared from five different natural occurring triterpenoids including oleanolic, ursolic, maslinic, betulinic, and platanic acid. All of which were derived from ethylene diamine holding an additional substituent connected to the ethylene diamine group. These derivatives were evaluated regarding their inhibitory activity of the enzymes acetylcholinesterase (AChE) and butyrylcholinesterase (BChE) employing Ellman's assay. We further determined the type of inhibition and inhibition constants. Carboxamides derived from platanic acid have been shown to be potent and selective BChE inhibitors. Especially the mixed-type inhibitor (3β)-*N*-(2-pyrrolidin-1-ylethyl)-3-acetyloxy-20-oxo-30-norlupan-28-amide (**35**) showed a remarkably low K_i of 0.07 ± 0.01 μM (K_i' = 2.38 ± 0.48 μM) for the inhibition of BChE.

Keywords: acetylcholinesterase; butyrylcholinesterase; triterpenoids

1. Introduction

It is now more than a century since Alois Alzheimer, a German physician, described a new disease of the brain [1], being today one of the most threatening plagues for the elderly and one of the greatest challenges for chemists and biologists. The eponymous disease is today one of the greatest scourges of humanity. Alzheimer's disease (AD), the most common cause of dementia, is a neurodegenerative disorder, causing the progressive loss of cognitive functions and memory. As a result of demographic changes, the number of AD patients is steadily rising. This disease is characterized by an increasing decline in acetylcholine (ACh, neurotransmitter) levels in the cholinergic system [2]. A common strategy for the management of AD is to develop inhibitors that suppress the degradation of ACh caused by hydrolases acetylcholinesterase (AChE, EC 3.1.1.7) and butyrylcholinesterase (BChE, EC 3.1.1.8).

Ursolic acid (**UA**), oleanolic acid (**OA**), maslinic acid (**MA**), betulinic acid (**BA**), and platanic acid (**PA**) are naturally occurring pentacyclic triterpenoids (Figure 1), which are widely distributed in various plants. Triterpenes and their derivatives represent a group of pharmacologically interesting substances showing a variety of biological activities such as antitumor [3], antiviral, and anti-HIV [4,5], antibacterial [6,7], and antifungal [8] properties. Pentacyclic triterpenoic acids have already been tested for cholinergic activities with moderate anti-cholinesterase activity [9,10]. Cyclic terpene derivatives, which act as inhibitors of cholinesterases, have been our research focus for a long time. Results from our group have demonstrated that structural modifications of the terpenoid backbone have a high impact onto the inhibitory potential for cholinesterases (ChEs) [11–13]. Exceptionally good ChE inhibition has been found for some amino derivatives of platanic acid [12], as well as for amides of dehydroabietylamine [14] and 12-hydroxydehydroabietylamine [15].

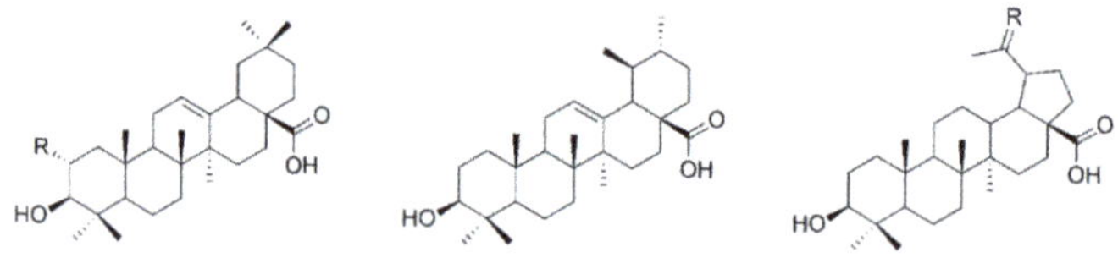

2 R = H (**OA**) 3 R = OH (**MA**) 1 (**UA**) 4 R = CH_2 (**BA**) 5 R = O (**PA**)

Figure 1. Structures of triterpenoic acids **1–5**.

In patients, BChE activity increases with progression of AD while the level of AChE remains constant [16]. Therefore, selective BChE inhibitors represent legitimate therapeutic options to improve the deficit in the neurotransmitter ACh. Deduced from the results of our previous studies, **UA**, **OA**, **MA**, **BA** and **PA** carboxamides holding a terminal amino moiety were synthesized from the parent pentacyclic triterpenoic acids. Altogether, 40 different compounds were synthesized and screened for their ChE inhibitory activity using Ellman's assay.

2. Results and Discussion

2.1. *Chemistry*

Acetylation of triterpenoic acids (Figure 1, **1–5**) furnished acetates **6–10** (Scheme 1). Reaction of **6–10** with oxalyl chloride and subsequent treatment with various amines derived from ethylene diamine gave carboxamides **11–15**, **21–25**, **31–35** and **41–45**. Their deacetylation with KOH/MeOH gave compounds **16–20**, **26–30**, **36–40** and **46–50**, respectively.

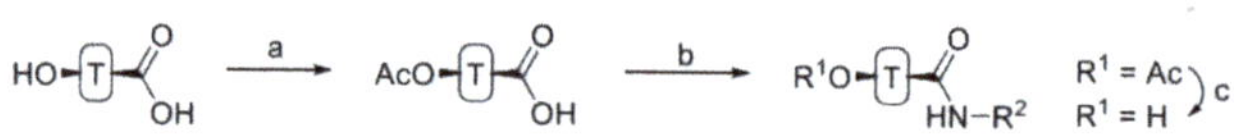

1-5 6-10 11-50

	R^2	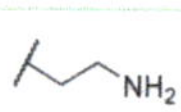		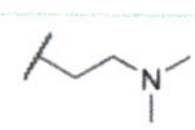		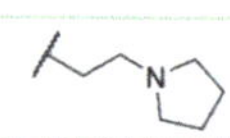		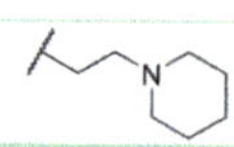	
	R^1	Ac	H	Ac	H	Ac	H	Ac	H
Backbone	**UA**	**11**	**16**	**21**	**26**	**31**	**36**	**41**	**46**
	OA	**12**	**17**	**22**	**27**	**32**	**37**	**42**	**47**
	MA	**13**	**18**	**23**	**28**	**33**	**38**	**43**	**48**
	BA	**14**	**19**	**24**	**29**	**34**	**39**	**44**	**49**
	PA	**15**	**20**	**25**	**30**	**35**	**40**	**45**	**50**

Scheme 1. Generalized representation of **1–5** and preparation of triterpenoic carboxamides **11–50**: (**a**) Ac_2O, DCM, TEA, 25 °C, 2 days, 90–96%; (**b**) oxalyl chloride, DCM, DMF, 0–25 °C, 1 h, then amine, 25 °C, 2 h; (**c**) MeOH/KOH, 25 °C, 2–3 days.

2.2. *Biology*

In this study, carboxamides with different triterpenoic backbones (ursolic, oleanolic, maslinic, betulinic, and platanic acids), and various amine residues (Scheme 1), were investigated. All compounds **11–50** (except those being not soluble under the conditions of the assay) were subjected to Ellman's assays to determine their inhibition rates and constants (K_i for competitive inhibition and K_i' for uncompetitive inhibition) for the cholinesterases AChE and BChE. Galantamine hydrobromide (**GH**), one of the gold standard drugs for treating AD symptoms, was used for comparison. In general, pre-screening the 37 triterpenoic acid amides using AChE (electric eel) and BChE (equine serum) identified seventeen compounds as potential inhibitors with inhibition rates equal or even better than **GH**. The results are summarized in Figures 2 and 3 (details are found in the Supplementary Part, Table S1).

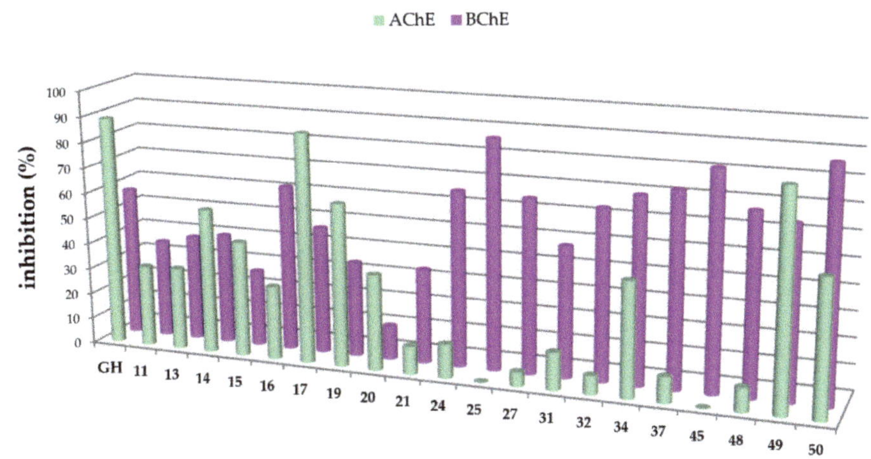

Figure 2. Percentage of inhibition of selected carboxamides and galantamine hydrobromide (**GH** as standard), final concentration of the inhibitor 10 µM, determined by Ellman's assay using acetylcholinesterase (green in the front) and butyrylcholinesterase (purple in the back).

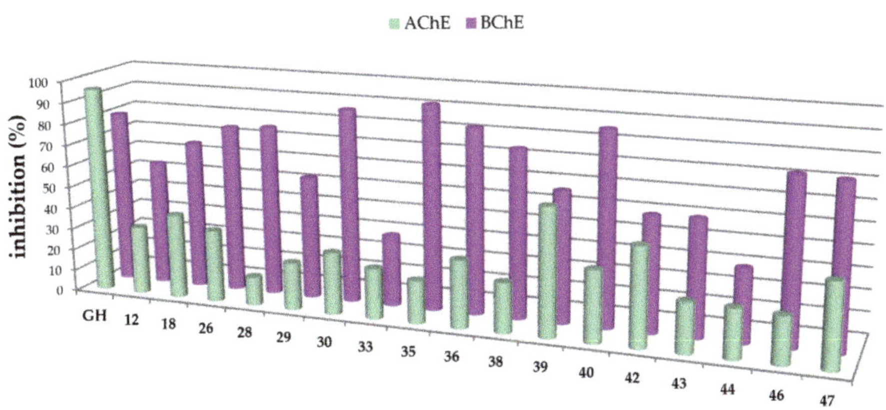

Figure 3. Percentage of inhibition of selected carboxamides and galantamine hydrobromide (**GH** as standard), final concentration of the inhibitor 30 µM, determined by Ellman's assay using acetylcholinesterase (green in the front) and butyrylcholinesterase (purple in the back).

The carboxamides showed a considerably higher inhibition for BChE than for AChE. With the exception of **17** (88.61 ± 0.22%) and **49** (82.72 ± 0.09%), none of the tested compounds showed any notable activity for AChE. Further kinetic studies (see Supplementary Part, Table S2) showed (3β)-*N*-(2-piperidin-1-ylethyl)-3-hydroxy-lup-20(29)-en-28-amide (**49**) as a good mixed-type AChE inhibitor with inhibition constants: K_i = 1.00 ± 0.09 µM and K_i' = 1.42 ± 0.08 µM. The results of the screening experiments suggest that many of the synthesized carboxylic acid amides inhibit BChE activity to a high degree. Fifteen derivatives showed promising inhibitory rates, five of which seemed to inhibit BChE almost completely (88.37–94.60% inhibition rate). The evaluation of the

Dixon [17], Cornish-Bowden [18], and Lineweaver-Burk [19] plots showed that all active compounds were mixed-type inhibitors with a dominating competitive part. The results from these measurements are compiled in Tables 1 and 2.

Table 1. Significant results of the BChE inhibition assay. Inhibitory constants (K_i and K_i' in μM), determined using Ellman's assays employing butyrylcholinesterase (BChE, equine serum) with galantamine hydrobromide (**GH**) as standard.

Compound	K_i in μM/K_i' in μM (Type of Inhibition)	Compound	K_i in μM/K_i' in μM (Type of Inhibition)
GH	2.30 ± 0.17 (competitive)	**32**	2.02 ± 0.56/29.88 ± 1.13 (mixed-type)
16	5.55 ± 0.12/12.15 ± 0.45 (mixed-type)	**34**	0.39 ± 0.04/> 5 (mixed-type)
17	1.93 ± 0.13/7.68 ± 0.61 (mixed-type)	**36**	0.47 ± 0.08/2.23 ± 0.12 (mixed-type)
21	6.48 ± 0.10/143.36 ± 0.02 (mixed-type)	**37**	0.55 ± 0.02/> 50 (mixed-type)
24	1.93 ± 0.13/7.68 ± 0.61 (mixed-type)	**38**	2.06 ± 0.03/14.15 ± 2.42 (mixed-type)
26	1.11 ± 0.09/3.10 ± 0.30 (mixed-type)	**39**	3.24 ± 0.58/> 9 (mixed-type)
27	1.31 ± 0.01/9.07 ± 0.02 (mixed-type)	**46**	2.37 ± 0.02/7.64 ± 0.22 (mixed-type)
28	2.20 ± 0.25/27.26 ± 3.47 (mixed-type)	**47**	1.57 ± 0.01/10.42 ± 1.36 (mixed-type)
29	3.52 ± 0.07/16.11 ± 0.18 (mixed-type)	**48**	2.26 ± 0.05/6.21 ± 0.47 (mixed-type)
31	4.42 ± 0.34/6.00 ± 0.47 (mixed-type)	**49**	2.52 ± 0.29/> 8 (mixed-type)

Table 2. Results of the BChE inhibition assay for platanic acid derivatives. Inhibitory constants (K_i and K_i' in μM), determined using Ellman's assay employing butyrylcholinesterase (BChE, equine serum) with galantamine hydrobromide (**GH**, K_i = 2.30 ± 0.17) as standard.

Compound	K_i in μM/K_i' in μM (Type of Inhibition)	Compound	K_i in μM/K_i' in μM (Type of Inhibition)
15	11.94 ± 1.17/17.14 ± 1.67 (mixed-type)	**35**	0.07 ± 0.01/2.38 ± 0.48 (mixed-type)
20	8.37 ± 0.67/> 110 (mixed-type)	**40**	0.45 ± 0.01/> 10 (mixed-type)
25	1.60 ± 0.13/1.60 ±0.11 (mixed-type)	**45**	0.71 ± 0.04/4.88 ± 0.21 (mixed-type)
30	1.12 ± 0.01/4.46 ± 0.01 (mixed-type)	**50**	0.97 ± 0.04/3.09 ± 0.03 (mixed-type)

The first set of compounds holding an aminoethyl residue, **11–14** and **16–19**, were moderate inhibitors of BChE, with oleanolic acid derivative **17** as the only noteworthy compound of this series (Ki = 1.93 ± 0.13 μM and K_i' = 7.68 ± 0.61 μM). The next set consisting of triterpenoic amides with a dimethylaminoethyl substituent (**21–24** and **26–29**) showed good inhibition with K_i values between 1.1 and 6.5 μM, respectively. For 2-piperidin-1-ylethyl substituted derivatives (**41–44** and **46–49**) inhibition constants in the same range were obtained. The group of 2-pyrrolidin-1-ylethyl substituted compounds (**31–34** and **36–39**) delivered the best results: derivatives **34** (from betulinic acid), **36** (ursolic acid backbone) and **37** (from oleanolic acid), held excellent K_i values of 0.39 ± 0.04 μM, 0.47 ± 0.08 μM, and 0.55 ± 0.02 μM, respectively. Compounds showing the highest selectivity (F, expressed by percent inhibition for AChE divided by percent inhibition for BChE) towards BChE were compounds **27** (F = 0.08) and **32** (F = 0.11).

Because of the known inhibitory effect [12], a set of eight platanic acid derivatives was investigated (Table 2). Compounds with an aminoethyl residue (**15** and **20**) were the weakest BChE inhibitors in our test. Compared to the standard **GH** (K_i = 2.30 ± 0.17 μM), 2-(dimethylamino)ethyl substituted derivatives showed good inhibition constants (K_i = 1.60 ± 0.13 μM for **25** and K_i = 1.12 ± 0.01 μM for **30**).

Compounds, carrying a piperidinyl group, showed inhibition even in nanomolar concentrations. For compound **45** (K_i = 0.71 ± 0.04 μM) and **50** (K_i = 0.97 ± 0.04 μM) excellent values were observed. As already described above, introducing a pyrrolidinyl moiety resulted in compounds with excellent K_i values. A very low inhibition constant (K_i = 0.45 ± 0.01 μM) was measured for compound **40**. The most active compound was (3β)-*N*-(2-pyrrolidin-1-ylethyl)-3-acetyloxy-20-oxo-30-norlupan-28-amide (**35**) showing inhibition constants of K_i = 0.07 ± 0.01 μM and K_i' = 2.38 ± 0.48 μM for BChE. This compound proved to be a selective (F ≈ 0.21) inhibitor for BChE. Except for **15** and **20**, all **PA**-derived compounds showed reasonably high BChE selectivity.

To explain the results from the biological testing, some molecular modeling studies were performed. Table 3 summarizes the results for the most favored docking position of each ligand. From these results, nice correlations between the gold fitness values (the higher the value the better the predicted affinity between ligand enzyme), as well as the interaction energies, with the experimental K_i values were established.

Table 3. Results of the docking studies with gold fitness values (PLP: Piecewise Linear Potential) and interaction energies (IE).

Compound	Gold Fitness (ChemPLP)	IE (kcal/mol)	K_i
31	90.0	−57.8	4.42
34	98.7	−60.4	0.39
35	99.0	−64.0	0.07

Figure 4 displays details of the docking arrangements of ligands with BChE. All ligands fit into the binding pocket—preferentially based on hydrophobic interactions (Figure 5) with the side chains of residues W79, W228, L283, Y329, and F326. The highest affinity was observed for **35** (Figure 4a). Thus a hydrogen bond of the acetyl group of **35** with the phenolic hydroxyl group of Y79 stabilizes the interaction with BChE. This hydrogen bond, however, is not present in compounds **31** and **34**. The modifications in ring E of **31** (six-membered ring) compared to compounds **34** and **35** (five-membered rings) slightly changed the docking arrangement (cf. Figure 4d) and caused reduced hydrophobic interactions in particular with W79.

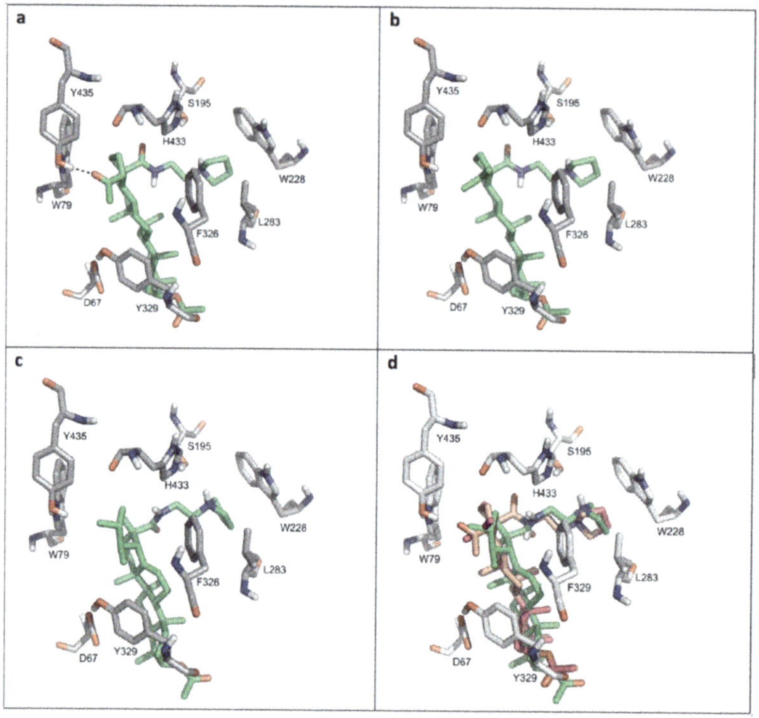

Figure 4. Details of the most favored docking arrangements of **35** (**a**), **34** (**b**) and **31** (**c**). (**d**) An overlay of all three ligands (orange colored: carbon atoms of **35**; magenta colored: carbon atoms of **34**; green colored: carbon atoms of **31**). Hydrophobic interactions with the side chains of W79, W228, L283, Y329, and F326 preferentially stabilize the docking poses. As for **35** (**a**) a hydrogen bond with Y435 is formed which might explain the higher affinity compared to the other compounds.

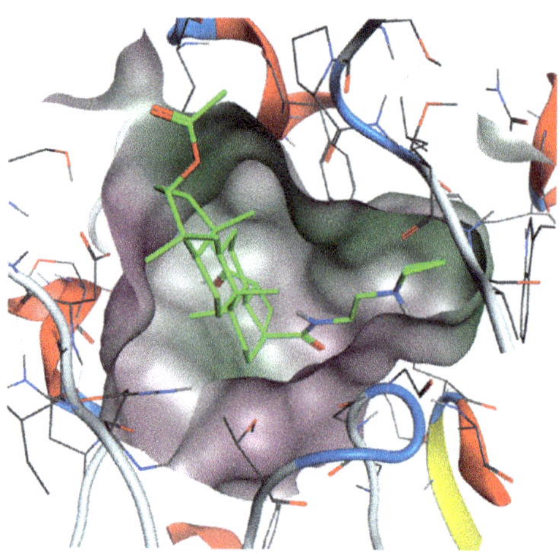

Figure 5. View from the top to the binding site of BChE with bound compound **35** (green colored: carbon atoms) with displayed lipophilic potential (green surface). The more hydrophilic site is colored magenta.

3. Materials and Methods

3.1. General

All chemicals, reagents and technical equipment were purchased in Germany unless otherwise stated. NMR spectra were recorded using the Varian spectrometers Gemini 2000 or Unity 500 (δ given in ppm, *J* in Hz; typical experiments: H-H-COSY, HMBC, HSQC, NOESY), MS spectra were taken on a Finnigan MAT LCQ 7000 (electrospray, voltage 4.1 kV, sheath gas nitrogen) instrument. The optical rotations were measured on a Perkin-Elmer polarimeter at 20 °C; TLC was performed on silica gel (Merck 5554, detection with cerium molybdate reagent) melting points were uncorrected (*Leica* hot stage microscope or BÜCHI Melting Point M-565) and elemental analyses were performed on a Foss-Heraeus Vario EL (CHNS) unit. IR spectra were recorded on a Perkin Elmer FT-IR spectrometer Spectrum 1000 or on a Perkin-Elmer Spectrum Two (UATR Two Unit). The solvents were dried according to usual procedures. The purity of the compounds was determined by HPLC and found to be >96%. Ursolic (**1**), betulinic (**4**), and platanic acid (**5**) were obtained from Betulinines (Stříbrná Skalice, Czech Republic), oleanolic acid (**2**) was purchased from Carbone Scientific (London, UK) and maslinic acid (**3**) was synthesized as previously described [20,21].

3.2. Biology

A TECAN SpectraFluorPlus working in the kinetic mode and measuring the absorbance at λ = 415 nm was used for the enzymatic studies. Acetylcholinesterase (from *Electrophorus electricus*), 5,5′-dithiobis-(2-nitrobenzoic acid) (DTNB) and acetylthiocholine iodide were purchased from Sigma. Butyrylcholinesterase (from equine serum) was purchased from Fluka. Preparation of the solutions, assay procedure, as well as molecular modeling conditions can be found in the Supplementary Materials.

4. Experimental

4.1. General

The preparation of the platanic carboxamides **15**, **20**, **25**, **30**, **35**, and **40** was performed as previously described [12,22]. Betulinic and ursolic carboxamides **11**, **14**, **16**, **19**, **21**, **24**, **26**, **29**, **31**, **34**, **36**, **39**, **41**, **44**, **46**, and **49** were prepared as previously reported in the literature [23]. Experimental procedures and full analytical data of these compounds can be found in the Supplementary Materials.

4.2. General Procedure A for the Acetylation of Triterpenoic Acids (**6–10**)

To a solution of triterpenoic acid **1–5** (11 mmol) in dry DCM (150 mL) was added triethylamine (4.6 mL, 33 mmol), acetic anhydride (3.1 mL, 33 mmol) and DMAP (cat.). After stirring for two days at 25 °C, a saturated solution of NH_3 in MeOH was added (3 mL) and the mixture was stirred for another 30 min. Dilution with DCM and subsequent aqueous work-up provided crude acetates. Recrystallization from EtOH yielded acetates **6–10** (90–96%) as colorless solids, whose spectroscopic data were in full agreement with data from the literature.

4.3. General Procedure B for the Synthesis of Triterpenoic Amides (**11–15**, **21–25**, **31–35**, and **41–45**)

Triterpenoic acetates **6–10** (0.8 mmol) were each dissolved in dry DCM (15 mL), cooled to 0 °C, then oxalyl chloride (3.2 mmol) and dry DMF (3 drops) were added. After warming to 25 °C, the mixture was stirred for 1 h. The solvent was removed under reduced pressure, re-evaporated with dry THF (4 × 15 mL), and the residue was immediately resolved in dry DCM (10 mL). This solution was then added dropwise to a solution of the amino compound (2.4 mmol) in dry DCM (5 mL) and stirred at 25 °C for 2 h. After usual aqueous work-up, the solvent was removed under reduced pressure and the crude products were subjected to column chromatography (silica gel, chloroform/methanol mixtures). Compounds **11–15**, **21–25**, **31–35**, and **41–45** were each obtained as colorless solids.

4.4. *General Procedure C for the Deacetylation of Triterpenoic Amides* (**16–20**, **26–30**, **36–40**, *and* **46–50**)

To a solution of the acetylated amide (0.33 mmol) in methanol (10 mL) was added a solution of potassium hydroxide (1.65 mmol) in methanol (2 mL). The mixture was stirred at 25 °C for 2 or 3 days. After completion of the reaction (as indicated by TLC), aq. HCl was added until pH = 7. After usual work-up, the solvent was removed under reduced pressure and the residue was subjected to column chromatography (silica gel, chloroform/methanol mixtures) yielding compounds **16–20**, **26–30**, **36–40**, and **46–50** each as colorless solids. *(3β)-3-Acetyloxy-urs-12-en-28-oic acid* (**6**), Compound **6** was prepared according to general procedure A from ursolic acid (**1**). Yield: 96%; m.p. 287–290 °C (lit.: 289–290 °C [24]).

(3β)-3-Acetyloxy-olean-12-en-28-oic acid (**7**), Compound **7** was prepared according to general procedure A from oleanolic acid (**2**). Yield: 90%; m.p. 259–261 °C (lit.: 255–257 °C [25]).

(2α,3β)-2,3-Diacetyloxy-olean-12-en-28-oic acid (**8**), Compound **8** was prepared according to general procedure A from maslinic acid (**3**). Yield: 91%; m.p. 172–175 °C (l.: 170–173 °C [26]).

(3β)-3-Acetyloxy-lup-20(29)en-28-oic acid (**9**), Compound **9** was prepared according to general procedure A from betulinic acid (**4**). Yield: 93%; m.p. 281–284 °C (lit.: 280–282 °C [27]).

(3β)-3-Acetyloxy-20-oxo-30-norlupan-28-oic acid (**10**), Compound **10** was prepared according to general procedure A from platanic acid (**5**). Yield: 94%; m.p. 256–259 °C (lit.: 252–255 °C [28]).

(3β)-N-(2-Aminoethyl)-3-acetyloxy-olean-12-en-28-amide (**12**), Compound **12** was prepared from **7** according to general procedure B using ethylenediamine as amino compound. Column chromatography (SiO_2, $CHCl_3$/MeOH 9:1) gave **12** (yield: 75%); m.p. 212–215 °C (decomp.); $[\alpha]_D$ = +37.8° (c 0.350, $CHCl_3$); R_f = 0.67 ($CHCl_3$/MeOH/NH_4OH 90:10:1); IR (ATR): ν = 2944 m, 1732 m, 1628 m, 1523 m, 1364 s, 1244 s, 1027 m, 985 m, 824 m, 752 m cm^{-1}; ^{1}H-NMR (400 MHz, $CDCl_3$): δ = 7.04 (t, J = 5.5 Hz, 1H, NH), 5.40 (t, J = 3.4 Hz, 1H, 12-H), 4.53–4.45 (m, 1H, 3-H), 3.68–3.56 (m, 1H, 31-H_a), 3.40–3.29 (m, 1H, 31-H_b), 3.24–3.11 (m, 2H, 32-H), 2.65 (dd, J = 12.7, 4.6 Hz, 1H, 18-H), 2.04 (s, 3H, Ac), 2.01–1.82 (m, 3H, 16-H_a, 11-H_a, 11-H_b), 1.79–1.22 (m, 14H, 19-H_a, 1-H_a, 2-H_a, 2-H_b, 7-H_a, 7-H_b, 9-H, 16-H_b, 6-H_a, 15-H_a, 22-H_a, 6-H_b, 21-H_a, 22-H_b), 1.22–1.11 (m, 2H, 19-H_b, 21-H_b), 1.14 (s, 3H, 27-H), 1.10–0.95 (m, 2H, 1-H_b, 15-H_b), 0.93 (s, 3H, 25-H), 0.90 (s, 3H, 30-H), 0.89 (s, 3H, 29-H), 0.86 (s, 3H, 23-H), 0.85 (s, 3H, 24-H), 0.84–0.79 (m, 1H, 5-H), 0.73 (s, 3H, 26-H) ppm; ^{13}C-NMR (101 MHz, $CDCl_3$): δ = 180.8 (C-28), 171.0 (Ac), 144.0 (C-13), 123.1 (C-12), 80.7 (C-3), 55.2 (C-5), 47.5 (C-9), 46.6 (C-19), 46.3 (C-17), 41.8 (C-14), 41.4 (C-18), 40.4 (C-32), 39.4 (C-8), 38.2 (C-1), 38.1 (C-31), 37.7 (C-4), 36.9 (C-10), 34.2 (C-21), 33.1 (C-30), 32.3 (C-22), 32.2 (C-7), 30.6 (C-20), 28.0 (C-23), 27.2 (C-15), 25.8 (C-27), 23.6 (C-2), 23.5 (C-16), 23.5 (C-11), 23.5 (C-29), 21.3 (Ac), 18.2 (C-6), 16.9 (C-26), 16.7 (C-24), 15.5 (C-25) ppm; MS (ESI, MeOH): m/z = 541.3 (100%, $[M + H]^+$); analysis calcd. for $C_{34}H_{56}N_2O_3$ (540.83): C 75.51, H 10.44, N 5.18; found: C 75.42, H 10.57, N 5.07.

(2α,3β)-N-(2-Aminoethyl)-2,3-diacetyloxy-olean-12-en-28-amide (**13**), Compound **13** was prepared from **8** according to general procedure B using ethylenediamine as amino compound. Column chromatography (SiO_2, $CHCl_3$/MeOH 9:1) gave **13** (yield: 74%); m.p. 151–154 °C; $[\alpha]_D$ = +18.7° (c 0.330, $CHCl_3$); R_f = 0.63 ($CHCl_3$/MeOH/NH_4OH 90:10:1); IR (KBr): ν = 3426 br s, 2946 s, 1742 s, 1636m, 1522m, 1458m, 1436w, 1368m, 1254s, 1044m cm^{-1}; ^{1}H-NMR (400 MHz, $CDCl_3$): δ = 6.36 (t, J = 5.5 Hz, 1H, NH), 5.37 (t, J = 3.6 Hz, 1H, 12-H), 5.08 (ddd, J = 11.1, 10.9, 4.6 Hz, 1H, 2-H), 4.73 (d, J = 10.3 Hz, 1H, 3-H), 3.48–3.39 (m, 1H, 31-H_a), 3.12–3.02 (m, 1H, 31-H_b), 2.87–2.76 (m, 2H, 32-H), 2.56 (dd, J = 13.1, 4.3 Hz, 1H, 18-H), 2.04 (s, 3H, Ac), 2.08–1.83 (m, 4H, 1-H_a, 16-H_a, 11-H_a, 11-H_b), 1.97 (s, 3H, Ac), 1.80–1.24 (m, 11H, 19-H_a, 22-H_a, 16-H_b, 9-H, 22-H_b, 6-H_a, 15-H_a, 7-H_a, 6-H_b, 21-H_a, 7-H_b), 1.23–1.09 (m, 2H, 21-H_b, 19-H_b), 1.14 (s, 3H, 27-H), 1.11–0.99 (m, 2H, 1-H_b, 15-H_b), 1.04 (s, 3H, 25-H), 0.99–0.92 (m, 1H, 5-H), 0.90 (s, 9H, 24-H, 29-H, 30-H), 0.89 (s, 3H, 23-H), 0.76 (s, 3H, 26-H) ppm; ^{13}C-NMR (101 MHz, $CDCl_3$): δ = 178.8 (C-28), 170.9 (Ac), 170.7 (Ac), 144.9 (C-13), 122.5 (C-12), 80.7 (C-3), 70.1 (C-2), 54.9 (C-5), 47.6 (C-9), 46.8 (C-19), 46.5 (C-17), 44.0 (C-1), 42.3 (C-18), 42.2 (C-14), 41.8

(C-31), 41.4 (C-32), 39.6 (C-8), 39.5 (C-4), 38.2 (C-10), 34.3 (C-21), 33.1 (C-30), 32.9 (C-22), 32.3 (C-7), 30.9 (C-20), 28.5 (C-23), 27.4 (C-15), 25.9 (C-27), 23.8 (C-16), 23.8 (C-29), 23.7 (C-11), 21.3 (Ac), 21.0 (Ac), 18.3 (C-6), 17.8 (C-24), 17.1 (C-26), 16.6 (C-25) ppm; MS (ESI, MeOH): m/z = 599 (100%, $[M + H]^+$); analysis calcd. for $C_{36}H_{58}N_2O_5$ (598.87): C 72.20, H 9.76, N 4.68; found: C 72.01, H 9.92, N 4.44.

(3β)-N-(2-Aminoethyl)-3-hydroxy-olean-12-en-28-amide (**17**), Compound **17** was prepared from **12** according to general procedure C. Column chromatography (SiO_2, $CHCl_3$/MeOH/NH_4OH 90:10:0.1) gave **17** (yield: 80%); m.p. 221–224 °C (decomp.); $[\alpha]_D$ = +55.3° (c 0.315, $CHCl_3$); R_f = 0.61 ($CHCl_3$/MeOH/NH_4OH 90:10:1); IR (ATR): ν = 3374w, 2941m, 1622m, 1530m, 1456m, 1387s, 1361s, 1344s, 1322s, 1187m, 1137m, 1097m, 1023m, 996m, 825m cm^{-1}; ^{1}H-NMR (500 MHz, CD_3OD): δ = 5.36 (t, J = 3.7 Hz, 1H, 12-H), 3.50–3.43 (m, 1H, 31-H_a), 3.37–3.32 (m, 1H, 31-H_b), 3.15 (dd, J = 11.3, 4.8 Hz, 1H, 3-H), 3.07–2.95 (m, 2H, 32-H), 2.79 (dd, J = 13.2, 4.4 Hz, 1H, 18-H), 2.10 (ddd, J = 13.8, 13.0, 4.0 Hz, 1H, 16-H_a), 1.99–1.86 (m, 2H, 11-H_a, 11-H_b), 1.83–1.74 (m, 1H, 19-H_a), 1.70–1.35 (m, 12H, 22-H_a, 1-H_a, 9-H, 15-H_a, 2-H_a, 2-H_b, 16-H_b, 6-H_a, 22-H_b, 7-H_a, 6-H_b, 21-H_a), 1.33–1.27 (m, 1H, 7-H_b), 1.25–1.14 (m, 2H, 21-H_b, 19-H_b), 1.18 (s, 3H, 27-H), 1.11–0.99 (m, 2H, 15-H_b, 1-H_b), 0.97 (s, 3H, 23-H), 0.96 (s, 3H, 29-H), 0.94 (s, 3H, 25-H), 0.92 (s, 3H, 30-H), 0.79 (s, 3H, 26-H), 0.78 (s, 3H, 24-H), 0.77–0.74 (m, 1H, 5-H) ppm; ^{13}C-NMR (126 MHz, CD_3OD): δ = 182.0 (C-28), 145.0 (C-13), 124.2 (C-12), 79.6 (C-3), 56.7 (C-5), 49.0 (C-9), 47.6 (C-17), 47.6 (C-19), 42.9 (C-14), 42.4 (C-18), 40.8 (C-32), 40.7 (C-8), 39.8 (C-4), 39.8 (C-1), 38.8 (C-31), 38.1 (C-10), 35.0 (C-21), 34.2 (C-22), 33.8 (C-7), 33.5 (C-30), 31.6 (C-20), 28.7 (C-23), 28.5 (C-15), 27.8 (C-2), 26.5 (C-27), 24.5 (C-11), 24.0 (C-29), 24.0 (C-16), 19.5 (C-6), 17.9 (C-26), 16.3 (C-24), 15.9 (C-25) ppm; MS (ESI, MeOH): m/z = 499.3 (100%, $[M + H]^+$); analysis calcd. for $C_{32}H_{54}N_2O_2$ (498.80): C 77.06, H 10.91, N 5.62; found: C 76.90, H 11.05, N 5.41.

(2α,3β)-N-(2-Aminoethyl)-2,3-dihydroxy-olean-12-en-28-amide (**18**), Compound **18** was prepared from **13** according to general procedure C. Column chromatography (SiO_2, $CHCl_3$/MeOH 9:1) gave **18** (yield: 60%); m.p. 260–266 °C (decompn.); $[\alpha]_D$ = +47.6° (c 0.335, $CHCl_3$); R_f = 0.36 ($CHCl_3$/MeOH/NH_4OH 90:10:1); IR (KBr): ν = 3424 br s, 944m, 1636m, 1528m, 1460w, 1384w, 1166w, 1050m cm^{-1}; ^{1}H-NMR (400 MHz, CD_3OD): δ = 5.36 (t, J = 3.7 Hz, 1H, 12-H), 3.61 (ddd, J = 11.3, 9.5, 4.5 Hz, 1H, 2-H), 3.43–3.35 (m, 1H, 31-H_a), 3.29–3.22 (m, 1H, 31-H_b), 2.90 (d, J = 9.6 Hz, 1H, 3-H), 2.89 (t, J = 6.5 Hz, 2H, 32-H), 2.81 (dd, J = 13.5, 4.5 Hz, 1H, 18-H), 2.14–2.03 (m, 1H, 16-H_a), 2.03–1.88 (m, 3H, 11-H_a, 11-H_b, 1-H_a), 1.78 (t, J = 13.5 Hz, 1H, 19-H_a), 1.70–1.26 (m, 10H, 9-H, 22-H_a, 15-H_a, 16-H_b, 6-H_a, 22-H_b, 7-H_a, 6-H_b, 21-H_a, 7-H_b), 1.23–1.13 (m, 2H, 19-H_b, 21-H_b), 1.18 (s, 3H, 27-H), 1.10–1.02 (m, 1H, 15-H_b), 1.01 (s, 3H, 23-H), 1.00 (s, 3H, 25-H), 0.95 (s, 3H, 30-H), 0.91 (s, 3H, 29-H), 0.93–0.89 (m, 1H, 1-H_b), 0.88–0.82 (m, 1H, 5-H), 0.80 (s, 3H, 24-H), 0.78 (s, 3H, 26-H) ppm; ^{13}C-NMR (101 MHz, CD_3OD): δ = 181.5 (C-28), 145.2 (C-13), 123.9 (C-12), 84.4 (C-3), 69.4 (C-2), 56.6 (C-5), 48.9 (C-9), 48.1 (C-1), 47.6 (C-17), 47.5 (C-19), 42.9 (C-14), 42.5 (C-18), 41.2 (C-32), 40.7 (C-8), 40.5 (C-4), 40.4 (C-31), 39.2 (C-10), 35.1 (C-21), 34.3 (C-22), 33.7 (C-7), 33.5 (C-30), 31.6 (C-20), 29.3 (C-23), 28.5 (C-15), 26.5 (C-27), 24.6 (C-11), 24.0 (C-29), 23.9 (C-16), 19.5 (C-6), 17.9 (C-26), 17.4 (C-24), 17.1 (C-25) ppm; MS (ESI, MeOH): m/z = 515 (100%, $[M + H]^+$); analysis calcd. for $C_{32}H_{54}N_2O_3$ (514.80): C 74.66, H 10.57, N 5.44; found: C 74.49, H 10.74, N 5.28.

(3β)-N-[2-(Dimethylamino)-ethyl]-3-acetyloxy-olean-12-en-28-amide (**22**), Compound **22** was prepared from **7** according to general procedure B using N,N-dimethylethylene diamine as amino compound. Column chromatography (SiO_2, $CHCl_3$/MeOH 95:5) gave **22** (yield: 95%); m.p. 227–230 °C (decomp.), Lit.: 255 °C (decomp.) [29]; $[\alpha]_D$ = +53.0° (c 0.320, $CHCl_3$), Lit.: +51.8° (c 0.34, $CHCl_3$)[29]; R_f = 0.38 ($CHCl_3$/MeOH 9:1); IR (ATR): ν = 2944m, 1732m, 1640w, 1464m, 1365m, 1244s, 1027m, 986m, 753m cm^{-1}; ^{1}H-NMR (400 MHz, $CDCl_3$): δ = 6.77 (t, J = 5.3 Hz, 1H, NH), 5.38 (t, J = 3.6 Hz, 1H, 12-H), 4.51–4.45 (m, 1H, 3-H), 3.61–3.51 (m, 1H, 31-H_a), 3.37–3.27 (m, 1H, 31-H_b), 2.89–2.83 (m, 2H, 32-H), 2.65–2.56 (m, 1H, 18-H), 2.61 (s, 6H, 33-H, 33′-H), 2.03 (s, 3H, Ac), 2.02–1.93 (m, 1H, 16-H_a), 1.93–1.85 (m, 2H, 11-H_a, 11-H_b), 1.72 (t, J = 13.4 Hz, 1H, 19-H_a), 1.66–1.22 (m, 13H, 22-H_a, 1-H_a, 2-H_a, 2-H_b, 16-H_b, 9-H, 22-H_b, 6-H_a, 15-H_a, 7-H_a, 6-H_b, 21-H_a, 7-H_b), 1.21–1.14 (m, 2H, 19-H_b, 21-H_b), 1.14 (s, 3H, 27-H), 1.10–0.97 (m, 2H, 1-H_b, 15-H_b), 0.92 (s, 3H, 25-H), 0.90 (s, 3H, 29-H), 0.89 (s, 3H, 30-H),

0.85 (s, 3H, 23-H), 0.84 (s, 3H, 24-H), 0.83–0.79 (m, 1H, 5-H), 0.74 (s, 3H, 26-H) ppm; ^{13}C-NMR (101 MHz, $CDCl_3$): δ = 179.3 (C-28), 171.1 (Ac), 144.2 (C-13), 123.1 (C-12), 81.0 (C-3), 57.8 (C-32), 55.4 (C-5), 47.6 (C-9), 46.6 (C-19), 46.5 (C-17), 44.7 (C-33, C-33′), 42.0 (C-14), 41.9 (C-18), 39.6 (C-8), 38.3 (C-1), 37.8 (C-4), 37.0 (C-10), 36.1 (C-31), 34.3 (C-21), 33.2 (C-30), 32.9 (C-22), 32.5 (C-7), 30.8 (C-20), 28.2 (C-23), 27.5 (C-15), 25.9 (C-27), 23.7 (C-29), 23.6 (C-16, C-2, C-11), 21.4 (Ac), 18.3 (C-6), 17.1 (C-24), 16.8 (C-26), 15.6 (C-25) ppm; MS (ESI, MeOH): m/z = 569.5 (100%, $[M + H]^+$); analysis calcd. for $C_{36}H_{60}N_2O_3$ (568.89): C 76.01, H 10.63, N 4.92; found: C 75.86, H 10.83, N 4.77.

(2α,3β)-N-[2-(Dimethylamino)ethyl]-2,3-diacetyloxy-olean-12-en-28-amide (**23**), Compound **23** was prepared from **8** according to general procedure B using N,N-dimethylethylene diamine as amino compound. Column chromatography (SiO_2, $CHCl_3$/MeOH 9:1) gave **23** (yield: 94%); m.p. 131–136 °C; $[\alpha]_D$ = +17.2° (c 0.340, $CHCl_3$); R_f = 0.31 ($CHCl_3$/MeOH 95:5); IR (KBr): ν = 3426br s, 2946s, 2772w, 1744s, 1654m, 1512m, 1462m, 1368m, 1252s, 1156w, 1044m cm^{-1}; ^{1}H-NMR (400 MHz, $CDCl_3$): δ = 6.51 (t, J = 4.4 Hz, 1H, NH), 5.33 (t, J = 3.6 Hz, 1H, 12-H), 5.08 (ddd, J = 11.2, 10.3, 4.6 Hz, 1H, 2-H), 4.75 (d, J = 10.3 Hz, 1H, 3-H), 3.39–3.29 (m, 1H, 31-H_a), 3.18–3.09 (m, 1H, 31-H_b), 2.56–2.48 (m, 1H, 18-H), 2.36 (t, J = 6.0 Hz, 2H, 32-H), 2.21 (s, 6H, 33-H, 33′-H), 2.05 (s, 3H. Ac), 2.04–1.99 (m, 1H, 1-H_a), 1.97 (s, 3H, Ac), 1.96–1.87 (m, 3H, 16-H_a, 11-H_a, 11-H_b), 1.81–1.25 (m, 11H, 19-H_a, 22-H_a, 16-H_b, 9-H, 22-H_b, 15-H_a, 6-H_a, 7-H_a, 6-H_b, 21-H_a, 7-H_b), 1.22–1.12 (m, 2H, 21-H_b, 19-H_b), 1.14 (s, 3H, 27-H), 1.05 (s, 3H, 25-H), 1.10–0.99 (m, 2H, 1-H_b, 15-H_b), 0.99–0.94 (m, 1H, 5-H), 0.90 (s, 9H, 24-H, 29-H, 30-H), 0.89 (s, 3H, 23-H), 0.79 (s, 3H, 26-H) ppm; ^{13}C-NMR (101 MHz, $CDCl_3$): δ = 178.1 (C-28), 170.9 (Ac), 170.7 (Ac), 144.7 (C-13), 122.5 (C-12), 80.7 (C-3), 70.2 (C-2), 57.7 (C-32), 55.0 (C-5), 47.6 (C-9), 46.8 (C-19), 46.5 (C-17), 45.4 (C-33, C-33′), 44.1 (C-1), 42.5 (C-18), 42.2 (C-14), 39.6 (C-8), 39.5 (C-4), 38.2 (C-10), 37.0 (C-31), 34.3 (C-21), 33.2 (C-30), 32.8 (C-22), 32.5 (C-7), 30.9 (C-20), 28.6 (C-23), 27.5 (C-15), 25.8 (C-27), 23.8 (C-16), 23.7 (C-11), 23.7 (C-29), 21.3 (Ac), 21.0 (Ac), 18.4 (C-6), 17.8 (C-24), 17.0 (C-26), 16.7 (C-25) ppm; MS (ESI, MeOH): m/z = 627 (100%, $[M + H]^+$); analysis calcd. for $C_{38}H_{62}N_2O_5$ (626.92): C 72.80, H 9.97, N 4.47; found: C 72.63, H 10.14, N4.29.

(3β)-N-[2-(Dimethylamino)-ethyl]-3-hydroxy-olean-12-en-28-amide (**27**), Compound **27** was prepared from **22** according to general procedure C. Column chromatography (SiO_2, $CHCl_3$/MeOH 95:5) gave **27** (yield: 85%); m.p. 187–190 °C; $[\alpha]_D$ = +42.6° (c 0.320, MeOH); R_f = 0.29 ($CHCl_3$/MeOH 9:1); IR (ATR): ν = 3383br w, 2943m; 1624m, 1532m, 1431s, 1387s, 1319s, 1186m, 1032m, 996m, 823m, 752s cm^{-1}; ^{1}H-NMR (500 MHz, $CDCl_3$): δ = 6.73 (t, J = 5.2 Hz, 1H, NH), 5.40 (t, J = 3.6 Hz, 1H, 12-H), 3.63–3.53 (m, 1H, 31-H_a), 3.35–3.27 (m, 1H, 31-H_b), 3.21 (dd, J = 11.4, 4.3 Hz, 1H, 3-H), 2.87–2.81 (m, 2H, 32-H), 2.64–2.57 (m, 1H, 18-H), 2.60 (s, 6H, 33-H, 33′-H), 1.99 (ddd, J = 13.7, 13.6, 3.9 Hz, 1H, 16-H_a), 1.93–1.89 (m, 2H, 11-H_a, 11-H_b), 1.73 (t, J = 13.4 Hz, 1H, 19-H_a), 1.68–1.30 (m, 12H, 22-H_a, 16-H_b, 1-H_a, 2-H_a, 2-H_b, 9-H, 22-H_b, 6-H_a, 15-H_a, 7-H_a, 6-H_b, 21-H_a), 1.26 (ddd, J = 12.3, 3.0, 2.9 Hz, 1H, 7-H_b), 1.22–1.14 (m, 2H, 21-H_b, 19-H_b), 1.15 (s, 3H, 27-H), 1.05 (ddd, J = 14.1, 3.6, 3.5 Hz, 1H, 15-H_b), 0.98 (s, 3H, 23-H), 0.98–0.93 (m, 1H, 1-H_b), 0.92 (s, 3H, 29-H), 0.90 (s, 3H, 25-H), 0.90 (s, 3H, 30-H), 0.78 (s, 3H, 24-H), 0.75 (s, 3H, 26-H), 0.74–0.71 (m, 1H, 5-H) ppm; ^{13}C-NMR (126 MHz, $CDCl_3$): δ = 179.2 (C-28), 144.2 (C-13), 123.2 (C-12), 79.1 (C-3), 57.8 (C-32), 55.3 (C-5), 47.7 (C-9), 46.7 (C-19), 46.6 (C-17), 44.7 (C-33, C-33′), 42.1 (C-14), 42.0 (C-18), 39.6 (C-8), 38.9 (C-4), 38.6 (C-1), 37.1 (C-10), 36.0 (C-31), 34.3 (C-21), 33.2 (C-30), 32.9 (C-22), 32.6 (C-7), 30.8 (C-20), 28.2 (C-23), 27.5 (C-15), 27.3 (C-2), 25.9 (C-27), 23.7 (C-16), 23.7 (C-29), 23.6 (C-11), 18.5 (C-6), 17.2 (C-26), 15.7 (C-24), 15.5 (C-25) ppm; MS (ESI, MeOH): m/z = 527.4 (100%, $[M + H]^+$); analysis calcd. for $C_{34}H_{58}N_2O_2$ (526.85): C 77.51, H 11.10, N 5.32; found: C 77.37, H 11.31, N 5.16.

(2α,3β)-N-[2-(Dimethylamino)-ethyl]-2,3-dihydroxy-olean-12-en-28-amide (**28**), Compound **28** was prepared from **23** according to general procedure C. Column chromatography (SiO_2, $CHCl_3$/MeOH 9:1) gave **28** (yield: 66%); m.p. 146–149 °C; $[\alpha]_D$ = +43.1° (c 0.355, $CHCl_3$); R_f = 0.10 ($CHCl_3$/MeOH 95:5); IR (KBr): ν = 3418br s, 2946s, 2862s, 2826m, 2778m, 1636s, 1522m, 1462s, 1386m, 1364m, 1266w, 1188w, 1158w, 1098w, 1050s cm^{-1}; ^{1}H-NMR (400 MHz, $CDCl_3$): δ = 6.56 (t, J = 4.9 Hz, 1H, NH), 5.35 (t, J = 3.6 Hz, 1H, 12-H), 3.68 (ddd, J = 11.3, 9.5, 4.5 Hz, 1H, 2-H), 3.42–3.32 (m, 1H, 31-H_a), 3.20–3.11

(m, 1H, 31-H_b), 2.99 (d, J = 9.5 Hz, 1H, 3-H), 2.57–2.50 (m, 1H, 18-H), 2.40 (t, J = 6.1 Hz, 2H, 32-H), 2.25 (s, 6H, 33-H, 33′-H), 2.01–1.90 (m, 4H, 1-H_a, 16-H_a, 11-H_a, 11-H_b), 1.80–1.24 (m, 11H, 19-H_a, 22-H_a, 16-H_b, 9-H, 22-H_b, 15-H_a, 6-H_a, 7-H_a, 6-H_b, 21-H_a, 7-H_b), 1.22–1.16 (m, 2H, 19-H_b, 21-H_b), 1.15 (s, 3H, 27-H), 1.06–0.99 (m, 1H, 15-H_b), 1.02 (s, 3H, 23-H), 0.97 (s, 3H, 25-H), 0.95–0.87 (m, 1H, 1-H_b), 0.90 (s, 6H, 29-H, 30-H), 0.86–0.82 (m, 1H, 5-H), 0.82 (s, 3H, 24-H), 0.78 (s, 3H, 26-H) ppm; ^{13}C-NMR (101 MHz, $CDCl_3$): δ = 178.3 (C-28), 144.6 (C-13), 122.8 (C-12), 84.0 (C-3), 67.0 (C-2), 57.8 (C-32), 55.4 (C-5), 47.7 (C-9), 46.8 (C-19), 46.6 (C-1), 46.5 (C-17), 45.3 (C-33, C-33′), 42.4 (C-18), 42.2 (C-14), 39.6 (C-8), 39.3 (C-4), 38.3 (C-10), 36.9 (C-31), 34.3 (C-21), 33.2 (C-30), 32.8 (C-22), 32.6 (C-7), 30.9 (C-20), 28.8 (C-23), 27.5 (C-15), 25.9 (C-27), 23.8 (C-16), 23.7 (C-11), 23.7 (C-29), 18.5 (C-6), 17.1 (C-24), 16.9 (C-26), 16.8 (C-25) ppm; MS (ESI, MeOH): m/z = 543 (100%, $[M + H]^+$); analysis calcd. for $C_{34}H_{58}N_2O_3$ (542.85): C 75.23, H 10.77, N 5.16; found: C 74.99, H 10.97, N 5.08.

(3β)-N-(2-Pyrrolidin-1-ylethyl)-3-acetyloxy-olean-12-en-28-amide (**32**), Compound **32** was prepared from **7** according to general procedure B using 1-(2-aminoethyl)-pyrrolidine as amino compound. Column chromatography (SiO_2, $CHCl_3$/MeOH 95:5) gave **32** (yield: 90%); m.p. 175–178 °C (decomp.), lit.: 258 °C (decomp.)[29]; $[\alpha]_D$ = +47.9° (c 0.315, $CHCl_3$), Lit.: +49.8° (c 0.350, $CHCl_3$)[29]; R_f = 0.39 ($CHCl_3$/MeOH 9:1); IR (ATR): ν = 2944m, 1731m, 1644w, 1522w, 1365m, 1244s, 1027m, 985m, 752m cm^{-1}; ^{1}H-NMR (400 MHz, $CDCl_3$): δ = 6.89 (t, J = 5.6 Hz, 1H, NH), 5.41 (t, J = 3.6 Hz, 1H, 12-H), 4.48 (dd, J = 10.3, 5.7 Hz, 1H, 3-H), 3.90–3.76 (m, 2H, 33-H_a, 33′-H_a), 3.73–3.64 (m, 1H, 31-H_a), 3.51–3.41 (m, 1H, 31-H_b), 3.35–3.18 (m, 2H, 32-H), 2.99–2.86 (m, 2H, 33-H_b, 33′-H_b), 2.64 (dd, J = 13.5, 4.6 Hz, 1H, 18-H), 2.22–1.95 (m, 5H, 34-H, 34′-H, 16-H_a), 2.03 (s, 3H, Ac), 1.94–1.76 (m, 2H, 11-H_a, 11-H_b), 1.70 (t, J = 13.5 Hz, 1H, 19-H_a), 1.65–1.28 (m, 12H, 1-H_a, 2-H_a, 2-H_b, 7-H_a, 7-H_b, 16-H_b, 9-H, 6-H_a, 15-H_a, 22-H_a, 6-H_b, 21-H_a), 1.28–1.15 (m, 3H, 22-H_b, 19-H_b, 21-H_b), 1.13 (s, 3H, 27-H), 1.09–0.97 (m, 2H, 1-H_b, 15-H_b), 0.92 (s, 3H, 25-H), 0.91 (s, 3H, 29-H), 0.89 (s, 3H, 30-H), 0.85 (s, 3H, 23-H), 0.84 (s, 3H, 24-H), 0.84–0.79 (m, 1H, 5-H), 0.71 (s, 3H, 26-H) ppm; ^{13}C-NMR (101 MHz, $CDCl_3$): δ = 179.9 (C-28), 171.1 (Ac), 143.9 (C-13), 123.3 (C-12), 81.0 (C-3), 55.3 (C-5), 55.1 (C-32), 54.8 (C-33, C-33′), 47.6 (C-9), 46.5 (C-19), 46.5 (C-17), 41.9 (C-14), 41.5 (C-18), 39.5 (C-8), 38.2 (C-1), 37.8 (C-4), 37.0 (C-10), 36.4 (C-31), 34.2 (C-21), 33.2 (C-30), 32.9 (C-7), 32.5 (C-22), 30.8 (C-20), 28.2 (C-23), 27.5 (C-15), 25.9 (C-27), 23.6 (C-2), 23.6 (C-29), 23.6 (C-11), 23.6 (C-16), 23.4 (C-34, C-34′), 21.4 (Ac), 18.3 (C-6), 17.2 (C-26), 16.8 (C-24), 15.5 (C-25) ppm; MS (ESI, MeOH): m/z = 595.5 (100%, $[M + H]^+$); analysis calcd. for $C_{38}H_{62}N_2O_3$ (594.93): C 76.72, H 10.50, N 4.71; found: C 76.51, H 10.73, N 4.69.

(2α,3β)-N-(2-Pyrrolidin-1-ylethyl)-2,3-diacetyloxy-olean-12-en-28-amide (**33**), Compound **33** was prepared from **8** according to general procedure B using 1-(2-aminoethyl)-pyrrolidine as amino compound. Column chromatography (SiO_2, $CHCl_3$/MeOH 9:1) gave **33** (yield: 96%); m.p. 143–146 °C (decomp.); $[\alpha]_D$ = +20.0° (c 0.345, $CHCl_3$); R_f = 0.30 ($CHCl_3$/MeOH 95:5); IR (KBr): ν = 3426br s, 2948s, 2802w, 1744s, 1654m, 1508m, 1460m, 1368m, 1250s, 1152w, 1044m cm^{-1}; ^{1}H-NMR (400 MHz, $CDCl_3$): δ = 6.58 (t, J = 4.3 Hz, 1H, NH), 5.30 (t, J = 3.6 Hz, 1H, 12-H), 5.08 (ddd, J = 11.5, 10.3, 4.7 Hz, 1H, 2-H), 4.74 (d, J = 10.3 Hz, 1H, 3-H), 3.46–3.37 (m, 1H, 31-H_a), 3.24–3.13 (m, 1H, 31-H_b), 2.67–2.44 (m, 7H, 32-H, 33-H, 33′-H, 18-H), 2.04 (s, 3H, Ac), 2.03–1.98 (m, 1H, 1-H_a), 1.97 (s, 3H, Ac), 1.96–1.83 (m, 3H, 11-H_a, 11-H_b, 16-H_a), 1.82–1.76 (m, 4H, 34-H, 34′-H), 1.75–1.23 (m, 11H, 19-H_a, 22-H_a, 16-H_b, 9-H, 22-H_b, 15-H_a, 6-H_a, 7-H_a, 6-H_b, 21-H_a, 7-H_b), 1.24–1.10 (m, 2H, 21-H_b, 19-H_b), 1.14 (s, 3H, 27-H), 1.05 (s, 3H, 25-H), 1.09–0.99 (m, 2H, 1-H_b, 15-H_b), 0.98–0.93 (m, 1H, 5-H), 0.90 (s, 9H, 24-H, 29-H, 30-H), 0.89 (s, 3H, 23-H), 0.77 (s, 3H, 26-H) ppm; ^{13}C-NMR (101 MHz, $CDCl_3$): δ = 177.9 (C-28), 170.8 (Ac), 170.5 (Ac), 144.6 (C-13), 122.3 (C-12), 80.5 (C-3), 70.0 (C-2), 54.8 (C-5), 54.2 (C-32), 53.9 (C-33, C-33′), 47.5 (C-9), 46.7 (C-19), 46.3 (C-17), 43.9 (C-1), 42.3 (C-18), 42.0 (C-14), 39.4 (C-8), 39.3 (C-4), 38.1 (C-10), 37.8 (C-31), 34.2 (C-21), 33.0 (C-30), 32.6 (C-22), 32.3 (C-7), 30.7 (C-20), 28.4 (C-23), 27.3 (C-15), 25.6 (C-27), 23.7 (C-11), 23.7 (C-34, C-34′), 23.6 (C-16), 23.5 (C-29), 21.1 (Ac), 20.9 (Ac), 18.2 (C-6), 17.6 (C-24), 16.8 (C-26), 16.5 (C-25) ppm; MS (ESI, MeOH): m/z = 653 (100%, $[M + H]^+$); analysis calcd. for $C_{40}H_{64}N_2O_5$ (652.96): C 73.58, H 9.88, N 4.29; found: C 73.27, H 10.02, N 4.02.

(3β)-N-(2-Pyrrolidin-1-ylethyl)-3-hydroxy-olean-12-en-28-amide (**37**), Compound **37** was prepared from **32** according to general procedure C. Column chromatography (SiO_2, $CHCl_3$/MeOH 95:5) gave **37** (yield: 76%); m.p. 189–192 °C; $[\alpha]_D$ = +40.8° (c 0.325, MeOH); R_f = 0.28 ($CHCl_3$/MeOH 9:1); IR (ATR): ν = 3386br w, 2941m, 1631m, 1527m, 1320s, 1031m, 996m, 823m, 752m cm^{-1}; ^{1}H-NMR (500 MHz, $CDCl_3$): δ = 6.94 (t, J = 5.6 Hz, 1H, NH), 5.42 (t, J = 3.7 Hz, 1H, 12-H), 3.88–3.78 (m, 2H, 33-H_a, 33′-H_a), 3.73–3.62 (m, 1H, 31-H_a), 3.51–3.41 (m, 1H, 31-H_b), 3.35–3.22 (m, 2H, 33-H_b, 33′-H_b), 3.20 (dd, J = 11.5, 4.4 Hz, 1H, 3-H), 2.99–2.88 (m, 2H, 32-H), 2.66 (dd, J = 12.9, 3.7 Hz, 1H, 18-H), 2.19–2.05 (m, 4H, 34-H, 34′-H), 2.01 (ddd, J = 13.9, 13.8, 3.9 Hz, 1H, 16-H_a), 1.97–1.84 (m, 2H, 11-H_a, 11-H_b), 1.70 (t, J = 13.4 Hz, 1H, 19-H_a), 1.65–1.29 (m, 12H, 1-H_a, 16-H_b, 2-H_a, 2-H_b, 22-H_a, 22-H_b, 9-H, 6-H_a, 15-H_a, 7-H_a, 6-H_b, 21-H_a), 1.28–1.23 (m, 1H, 7-H_b), 1.21–1.15 (m, 2H, 19-H_b, 21-H_b), 1.14 (s, 3H, 27-H), 1.04 (ddd, J = 14.4, 3.5, 3.4 Hz, 1H, 15-H_b), 0.98 (s, 3H, 23-H), 0.97–0.92 (m, 1H, 1-H_b), 0.91 (s, 3H, 29-H), 0.90 (s, 3H, 25-H), 0.89 (s, 3H, 30-H), 0.77 (s, 3H, 24-H), 0.74–0.70 (m, 1H, 5-H), 0.71 (s, 3H, 26-H) ppm; ^{13}C-NMR (126 MHz, $CDCl_3$): δ = 180.1 (C-28), 143.8 (C-13), 123.4 (C-12), 79.1 (C-3), 55.3 (C-5), 55.1 (C-33, C-33′), 54.9 (C-32), 47.7 (C-9), 46.5 (C-17, C-19), 41.9 (C-14), 41.5 (C-18), 39.6 (C-8), 38.9 (C-4), 38.6 (C-1), 37.1 (C-10), 36.4 (C-31), 34.3 (C-21), 33.2 (C-30), 32.9 (C-22), 32.6 (C-7), 30.8 (C-20), 28.2 (C-23), 27.5 (C-15), 27.3 (C-2), 26.0 (C-27), 23.7 (C-29), 23.6 (C-16), 23.5 (C-11), 23.4 (C-34, C-34′), 18.5 (C-6), 17.2 (C-26), 15.7 (C-24), 15.5 (C-25) ppm; MS (ESI, MeOH): m/z = 553.5 (100%, $[M + H]^+$); analysis calcd. for $C_{36}H_{60}N_2O_2$ (552.89): C 78.21, H 10.94, N 5.07; found: C 78.01, H 11.13, N 4.83.

(2α,3β)-N-(2-Pyrrolidin-1-ylethyl)-2,3-dihydroxy-olean-12-en-28-amide (**38**), Compound **38** was prepared from **33** according to general procedure C. Column chromatography (SiO_2, $CHCl_3$/MeOH 9:1) gave **38** (yield: 67%); m.p. 153–156 °C (decomp.); $[\alpha]_D$ = +44.3° (c 0.330, $CHCl_3$); R_f = 0.10 ($CHCl_3$/MeOH 95:5); IR (KBr): ν = 3406br s, 2946s, 2878s, 2808m, 1636s, 1522m, 1462m, 1386m, 1364m, 1268w, 1194w, 1150w, 1050m cm^{-1}; ^{1}H-NMR (400 MHz, $CDCl_3$): δ = 6.72 (t, J = 5.1 Hz, 1H, NH), 5.34 (t, J = 3.6 Hz, 1H, 12-H), 3.67 (ddd, J = 11.3, 9.4, 4.4 Hz, 1H, 2-H), 3.53–3.43 (m, 1H, 31-H_a), 3.31–3.22 (m, 1H, 31-H_b), 2.99 (d, J = 9.5 Hz, 1H, 3-H), 2.80–2.63 (m, 6H, 32-H, 33-H, 33′-H), 2.56 (dd, J = 13.4, 4.3 Hz, 1H, 18-H), 2.02–1.89 (m, 4H, 1-H_a, 11-H_a, 11-H_b, 16-H_a), 1.89–1.83 (m, 4H, 34-H, 34′-H), 1.79–1.24 (m, 11H, 19-H_a, 22-H_a, 16-H_b, 9-H, 22-H_b, 6-H_a, 15-H_a, 7-H_a, 6-H_b, 21-H_a, 7-H_b), 1.23–1.11 (m, 2H, 21-H_b, 19-H_b), 1.14 (s, 3H, 27-H), 1.08–0.99 (m, 1H, 15-H_b), 1.02 (s, 3H, 23-H), 0.97 (s, 3H, 25-H), 0.94–0.87 (m, 1H, 1-H_b), 0.90 (s, 3H, 29-H), 0.89 (s, 3H, 30-H), 0.85–0.79 (m, 1H, 5-H), 0.81 (s, 3H, 24-H), 0.75 (s, 3H, 26-H) ppm; ^{13}C-NMR (101 MHz, $CDCl_3$): δ = 178.6 (C-28), 144.5 (C-13), 122.8 (C-12), 84.0 (C-3), 68.9 (C-2), 55.3 (C-5), 54.6 (C-32), 54.2 (C-33, C-33′), 47.7 (C-9), 46.8 (C-19), 46.5 (C-1), 46.5 (C-17), 42.2 (C-18), 42.1 (C-14), 39.6 (C-8), 39.3 (C-4), 38.3 (C-10), 37.6 (C-31), 34.3 (C-21), 33.2 (C-30), 32.8 (C-22), 32.5 (C-7), 30.8 (C-20), 28.8 (C-23), 27.5 (C-15), 25.9 (C-27), 23.8 (C-11), 23.7 (C-29), 23.7 (C-34, C-34′), 23.6 (C-16), 18.5 (C-6), 17.1 (C-26), 16.9 (C-24), 16.8 (C-25) ppm; MS (ESI, MeOH): m/z = 569 (100%, $[M + H]^+$); analysis calcd. for $C_{36}H_{60}N_2O_3$ (568.89): C 76.01, H 10.63, N 4.92; found: C 75.86, H 10.90, N 4.77.

(3β)-N-(2-Piperidin-1-ylethyl)-3-acetyloxy-olean-12-en-28-amide (**42**), Compound **42** was prepared from **7** according to general procedure B using 1-(2-aminoethyl)-piperidine as amino compound. Column chromatography (SiO_2, $CHCl_3$/MeOH 9:1) gave **42** (yield: 76%); m.p. = 163–166 °C (decomp.); $[\alpha]_D$ = +47.5° (c 0.305, $CHCl_3$); R_f = 0.55 (silica gel, $CHCl_3$/MeOH 9:1); IR (ATR): ν = 2943m, 2864w, 1732m, 1642w, 1523w, 1432m, 1388m, 1365s, 1330m, 1244vs, 1214m, 1149w, 1097w, 1027m, 1007m, 986m, 751m; ^{1}H-NMR (500 MHz, $CDCl_3$): δ = 6.93 (t, J = 5.3 Hz, 1H, NH), 5.42 (dd, J = 3.3 Hz, 3.3 Hz, 1H, 12-H), 4.51–4.44 (m, 1H, 3-H), 3.71 (dq, J = 11.9, 5.9 Hz, 1H, 31-H_a), 3.46 (dq, J = 12.2, 6.0 Hz, 1H, 31-H_b), 3.13 (s, 2H, 32-H), 2.64 (dd, J = 12.9, 3.3 Hz, 1H, 18-H), 2.03 (s, 3H, Ac), 2.02–1.84 (m, 8H, 2-H_a, 2-H_b, 16-H_a, 16-H_b. 33-H, 33′-H), 1.70 (m, 1H, 19-H_a), 1.66–1.40 (m, 15H, 1-H_a, 11-H_a, 22-H_a, 11-H_b, 9-H, 22-H_b, 6-H_a, 15-H_a, 7-H_a, 34-H, 34′-H, 35-H), 1.40–1.30 (m, 2H, 6-H_b, 21-H_a), 1.24 (m, 1H, 7-H_b), 1.22–1.14 (m, 2H, 19-H_b, 21-H_b), 1.13 (s, 3H, 27-H), 1.04 (m, 2H, 1-H_b, 15-H_b), 0.92 (s, 3H, 29-H) 0.92 (s, 3H, 26-H), 0.89 (s, 3H, 30-H), 0.85 (s, 3H, 23-H), 0.84 (s, 3H, 25-H), 0.81 (s, 1H, 5-H), 0.71 (s, 3H, 24-H); ^{13}C-NMR (126 MHz, $CDCl_3$): δ = 179.9 (C-28), 171.1 (Ac), 143.9 (C-13), 123.3 (C-12), 81.0 (C-3), 56.9 (C-32), 55.3 (C-5), 54.5 (C-33 + C-33′), 47.6 (C-9), 46.5 (C-19), 46.5 (C-17), 41.9 (C-14), 41.5

(C-18), 39.5 (C-8), 38.2 (C-1), 37.8 (C-4), 37.0 (C-10), 34.9 (C-31), 34.3 (C-21), 33.2 (C-30), 32.9 (C-22), 32.5 (C-7), 30.8 (C-20), 28.2 (C-23), 27.5 (C-15), 26.0 (C-27), 23.6 (C-2), 23.6 (C-29), 23.6 (C-11), 23.6 (C-16), 23.1 (C-34 + C-34′), 22.1 (C-35), 21.4 (Ac), 18.3 (C-6), 17.2 (C-24), 16.8 (C-25), 15.5 (C-26); MS (ESI, MeOH): m/z = 609.5 (100%, $[M + H]^+$); analysis calcd. for $C_{39}H_{64}N_2O_3$ (608.95): C 76.92, H 10.59, N 4.60; found: C 76.78, H 10.79, N 4.44.

(2α,3β)-N-(2-Piperidin-1-ylethyl)-2,3-diacetyloxy-olean-12-en-28-amide (**43**), Compound **43** was prepared from **8** according to general procedure B using 1-(2-aminoethyl)-piperidine as amino compound. Column chromatography (SiO_2, $CHCl_3$/MeOH 9:1) gave **43** (yield: 71%); m.p. = 130–133 °C (decomp.); $[\alpha]_D$ = +15.7° (c 0.35, $CHCl_3$); R_f = 0.47 (silica gel, $CHCl_3$/MeOH 9:1); IR (ATR): ν = 2938m, 2863w, 1741s, 1653m, 1508m, 1503m, 1466m, 1455m, 1443w, 1434w, 1367m, 1303w, 1247vs, 1229vs, 1155w, 1127w, 1041s, 1032s; ^{1}H-NMR (500 MHz, $CDCl_3$) δ = 6.69 (s, 1H, NH), 5.36 (dd, J = 3.4 Hz, 1H, 12-H), 5.08 (td, J = 11.4, 4.6 Hz, 1H, 2-H), 4.74 (d, J = 10.4 Hz, 1H, 3-H), 3.40 (m, 1H, 31-H_a), 3.20 (m, 1H, 31-H_b), 2.55 (d, J = 11.7 Hz, 1H, 18-H), 2.43 (s, 6 H, 32-H_a, 32-H_b, 33-H, 33′-H), 2.05 (s, 3H, Ac), 1.97 (s, 3H, Ac), 2.07–1.82 (m, 3H, 1-H_a, 16-H_a, 16-H_b), 1.78–1.42 (m, 15H, 19-H_a, 11-H_a, 22-H_a, 9-H, 11-H_b, 34-H, 34′-H, 6-H_a, 15-H_a, 22-H_b, 35-H, 7-H_a), 1.42–1.13 (m, 5H, 6-H_b, 21-H_a, 7-H_b, 21-H_b, 19-H_b), 1.14 (s, 3H, 27-H), 1.04 (s, 3H, 25-H), 1.09–0.92 (m, 3H, 1-H_b, 15-H_b, 5-H), 0.90 (s, 3H, 29-H), 0.90 (s, 3H, 24-H), 0.90 (s, 3H, 30-H), 0.90 (s, 3H, 23-H), 0.75 (s, 3H, 26-H); ^{13}C-NMR (126 MHz, $CDCl_3$) δ = 178.0, (C-28),170.9 (Ac), 170.7 (Ac), 144.8 (C-13), 122.4 (C-12), 80.7 (C-3), 70.2 (C-2), 57.0 (C-32), 55.0 (C-33 + C-33′), 54.4 (C-5), 47.6 (C-9), 46.9 (C-19), 46.4 (C-17), 44.0 (C-1), 42.3 (C-18), 42.1 (C-14), 39.6 (C-8), 39.5 (C-4), 38.3 (C-10), 35.8 (C-31), 34.3 (C-21), 33.2 (C-30), 32.8 (C-22), 32.3 (C-7), 30.9 (C-20), 28.6 (C-23), 27.4 (C-15), 25.8 (C-38 + C-38′), 25.9 (C-27), 24.2 (C-35), 23.7 (C-16 + C-11), 23.7 (C-29), 21.3 (Ac), 21.0 (Ac), 18.3 (C-6), 17.8 (C-24), 17.0 (C-26), 16.6 (C-25); MS (ESI, MeOH): m/z = 667.5 ($[M + H]^+$, 100%); analysis calcd. for $C_{41}H_{66}N_2O_5$ (666.99): C 73.83, H 9.97, N 4.20; found: C 73.69, H 10.18, N 4.01.

(3β)-N-(2-Piperidin-1-ylethyl)-3-acetyloxy-20-oxo-30-norlupan-28-amide (**45**), Compound **45** was prepared from **10** according to general procedure B using 1-(2-aminoethyl)-piperidine as amino compound. Column chromatography (SiO_2, $CHCl_3$/MeOH 9:1) gave **45** (yield: 63%); m.p. = 141–144 °C; $[\alpha]_D$ = −9.5° (c 0.34, $CHCl_3$); R_f = 0.43 (silica gel, $CHCl_3$/MeOH); IR (ATR): ν = 2939m, 2867w, 1733m, 1711m, 1657m, 1654m, 1517m, 1468m, 1451m, 1391w, 1367m, 1352m, 1316w, 1302w, 1243vs, 1196m, 1157m, 1130w, 1110w, 1090w, 1073w, 1026m, 979m; ^{1}H-NMR (500 MHz, $CDCl_3$) δ = 6.76 (s, 1H, NH), 4.45 (dd, J = 11.0, 5.1 Hz, 1H, 3-H), 3.40 (m, 4H, 18-H, 31-H_a, 31-H_b), 2.63–2.52 (m, 6H, 32-H_a, 32-H_b, 33-H, 33′-H), 2.21–2.14 (m, 1H, 13-H), 2.15 (s, 3H, 29-H), 2.09–2.00 (m, 3H, 9-H, 16-H_a, 21-H_a), 2.02 (s, 3H, Ac), 1.84–1.78 (dd, 1H, 1-H_a), 1.71–1.53 (m, 8H, 34-H, 34′-H, 19-H_a, 1-H_b, 16-H_b, 2-H_a), 1.53–1.20 (m, 14H, 13-H_a, 2-H_b, 21-H_b, 6-H_a, 35-H, 11-H_a, 12-H_a, 7-H_a, 7-H_b, 6-H_b, 11-H_b, 22-H_a, 22-H_b), 1.20–1.13 (m, 1H, 12-H_b), 1.10–1.00 (m, 2H, 15-H_a, 15-H_b), 0.98 (s, 3H, 27-H), 0.96–0.91 (m, 1H, 19-H_b), 0.88 (s, 3H, 26-H), 0.82 (s, 3H, 23-H), 0.82 (s, 3H, 25-H), 0.81 (s, 3H, 26-H), 0.79–0.75 (m, 1H, 5-H); ^{13}C-NMR (126 MHz, $CDCl_3$) δ = 212.9 (C-20), 176.5 (Ac), 171.0 (C-28), 81.0 (C-3), 57.3 (C-32), 55.8 (C-5), 55.5 (C-17), 54.3 (C-33), 51.4 (C-18), 50.5 (C-9), 50.1 (C-19), 42.4 (C-14), 40.8 (C-8), 38.5 (C-1), 38.0 (C-4), 37.9 (C-10), 37.2 (C-13), 37.0 (C-22), 35.3 (C-31), 34.4 (C-7), 32.9 (C-16), 30.3 (C-29), 29.6 (C-12), 28.8 (C-21), 28.1 (C-23), 27.3 (C-15), 25.3 (C-34), 23.8 (C-35), 23.8 (C-2), 21.4 (Ac), 21.1 (C-11), 18.3 (C-6), 16.6 (C-24), 16.3 (C-25), 16.3 (C-26), 14.8 (C-27); MS (ESI, MeOH): m/z = 611.5 (100%, $[M + H]^+$); analysis calcd. for $C_{38}H_{62}N_2O_4$ (610.92): C 74.71, H 10.23, N 4.59; found: C 74.56, H 10.51, N 4.39.

(3β)-N-(2-Piperidin-1-ylethyl)-3-hydroxy-olean-12-en-28-amide (**47**), Compound **47** was prepared from **42** according to general procedure C. Column chromatography (SiO_2, $CHCl_3$/MeOH 95:5) gave **47** (yield: 93%); m.p. = 184–186 °C (decomp.); $[\alpha]_D$ = +49.5° (c 0.365, $CHCl_3$); R_f = 0.26 (silica gel, $CHCl_3$/MeOH 9:1); IR (ATR): ν = 3372w, 2941m, 2864m, 1633m, 1527m, 1431s, 1387s, 1378s, 1357s, 1324vs, 1245m, 1212m, 1200w, 1188w, 1138w, 1093w, 1032m, 1006m, 997m, 752m; ^{1}H-NMR (500 MHz, $CDCl_3$) δ = 6.97 (t, J = 5.3 Hz, 1H, NH), 5.43 (dd, J = 3.4, 3.4 Hz, 1H, 12-H), 4.58 (s, 1H, OH), 3.71 (dq, J = 11.7, 5.8 Hz, 1H, 31-H_a), 3.46 (dq, J = 12.2, 5.9 Hz, 1H, 31-H_b), 3.22–3.17 (m, 1H, 3-H), 3.17–3.11 (m, 2H, 32-H_a, 32-H_b), 2.65 (dd, J = 13.4, 3.6 Hz, 2H, 18-H), 2.05–1.95 (m, 2H, 16-H_a, 16-H_b), 1.95–1.84

(m, 8H, 11-H_a, 11-H_b, 34-H, 34′-H, 35-H), 1.74–1.67 (m, 1H, 19-H_a), 1.65–1.45 (m, 10H, 1-H_a, 16-H_a, 22-H_a, 16-H_b, 22-H_b, 2-H_a, 2-H_b, 9-H, 6-H_a, 15-H_a), 1.45–1.39 (m, 1H, 7-H_a), 1.40–1.29 (m, 2H, 6-H_b, 21-H_a), 1.29–1.22 (m, 1H, 7-H_b), 1.21–1.15 (m, 2H, 19-H_b, 21-H_b), 1.14 (s, 3H, 27-H), 1.07–1.00 (m, 1H, 15-H_b), 0.97 (s, 3H, 23-H), 0.97–0.93 (m, 1H, 1-H_b), 0.92 (s, 3H, 29-H), 0.89 (s, 3H, 25-H), 0.89 (s, 3H, 30-H), 0.77 (s, 3H, 24-H), 0.73 (s, 1H, 5-H), 0.71 (s, 3H, 26-H); ^{13}C-NMR (126 MHz, $CDCl_3$) δ = 179.9 (C-28), 143.8 (C-13), 123.4 (C-12), 79.1 (C-3), 55.3 (C-32), 56.9 (C-5), 54.5 (C-33 + C-33′), 47.7 (C-9), 46.5 (C-19), 46.5 (C-17), 41.9 (C-14), 41.5 (C-18), 39.5 (C-8), 38.9 (C-4), 38.6 (C-1), 37.1 (C-10), 34.9 (C-31), 34.3 (C-21), 33.2 (C-30), 32.9 (C-22), 32.6 (C-7), 30.8 (C-20), 28.2 (C-23), 27.5 (C-15), 27.3 (C-2), 26.0 (C-27), 23.6 (C-29), 23.6 (C-16), 23.6 (C-11), 23.0 (C-34 + C-34′), 22.1 (C-35), 18.4 (C-6), 17.2 (C-26), 15.7 (C-24), 15.5 (C-25); MS (ESI, MeOH): m/z = 567.5 (100%, $[M + H]^+$); analysis calcd. for $C_{37}H_{62}N_2O_2$ (566.92): C 78.39, H 11.02, N 4.94; found: C 78.21,H 11.18, N 4.78.

(2α,3β)-N-(2-Piperidin-1-ylethyl)-2,3-dihydroxy-olean-12-en-28-amide (**48**), Compound **48** was prepared from **43** according to general procedure C. Column chromatography (SiO_2, $CHCl_3$/MeOH 9:1) gave **48** (yield: 87%); m.p. = 149–152 °C; $[\alpha]_D$ = +45.2° (c 0.31, $CHCl_3$); R_f = 0.21 (silica gel, $CHCl_3$/MeOH 9:1); IR (ATR): ν = 3378m, 2934vs, 2861m, 2853m, 1635s, 1513s, 1459s, 1454s, 1386s, 1378s, 1364m, 1348m, 1328w, 1303m, 1281m, 1266m, 1259m, 1232w, 1209w, 1194m, 1154m, 1129m, 1111w, 1096m, 1084m, 1050vs, 1016m, 992m, 958m, 660m; ^{1}H-NMR (500 MHz, $CDCl_3$) δ = 5.41 (s, 1H, 12-H), 3.74–3.64 (m, 1H, 2-H), 3.59–3.45 (m, 1H, 31-H_a), 3.43–3.25 (m, 1H, 31-H_b), 3.00 (d, J = 9.5 Hz, 1H, 3-H), 2.72–2.60 (m, 1H, 18-H), 2.64–2.36 (m, 2H, 32-H_a, 32-H_b), 2.02–1.92 (m, 4H, 1-H_a, 16-H_a, 16-H_b, 11-H_a), 1.77–1.49 (m, 17H, 1-H_b, 33-H, 33′-H, 9-H, 11-H_b, 34-H, 34′-H, 6-H_a, 22-H_a, 35-H, 7-H_a, 15-H_a), 1.50–1.42 (m, 1H, 7-H_b), 1.41–1.31 (m, 2H, 6-H_b, 21-H_a), 1.31–1.28 (m, 1H, 21-H_b), 1.19 (m, 3H, 22-H_b, 19-H_a), 1.15 (s, 3H, 27-H), 1.04 (m, 1H, 15-H_b), 1.03 (s, 3H, 23-H), 0.97 (s, 3H, 25-H), 0.93 (s, 3H, 29-H), 0.91 (s, 3H, 30-H), 0.92-0.88 (m, 1H, 19-H_b), 0.86-0.82 (m, 1H, 5-H), 0.82 (s, 3H, 26-H), 0.74 (s, 3H, 24-H); ^{13}C-NMR (126 MHz, $CDCl_3$) δ = 179.0 (C-28), 144.0 (C-13), 122.8 (C-12), 84.1 (C-3), 69.1 (C-2), 57.2 (C-32), 55.37 (C-33 + C-33′), 54.5 (C-5), 48.2 (C-17), 47.7 (C-9), 46.5 (C-19), 46.5 (C-1), 43.2 (C-18), 42.1 (C-14), 39.6 (C-8), 39.3 (C-4), 39.1 (C-10), 38.4 (C-31), 34.3 (C-21), 33.2 (C-30), 33.0 (C-22), 32.5 (C-7), 30.9 (C-20), 28.8 (C-23), 27.5 (C-15), 25.8 (C-34 + C-34′), 26.0 (C-27), 24.1 (C-35), 23.7 (C-16), 23.7 (C-11), 23.7 (C-29), 18.5 (C-6), 17.2 (C-24), 16.9 (C-26), 16.8 (C-25); MS (ESI, MeOH): m/z = 569.5 ($[M + H]^+$, 100%); analysis calcd. for $C_{37}H_{62}N_2O_3$ (582.91): C 76.24, H 10.72, N 4.81; found: C 75.97, H 10.93, N 4.57.

(3β)-N-(2-Piperidin-1-ylethyl)-3-hydroxy-20-oxo-30-norlupan-28-amide (**50**), Compound **50** was prepared from **45** according to general procedure C. Column chromatography (SiO_2, $CHCl_3$/MeOH 9:1) gave **50** (yield: 86%); m.p. = 153–156 °C (decomp.); $[\alpha]_D$ = −21.2° (c 0.115, $CHCl_3$); R_f = 0.16 (silica gel, $CHCl_3$/MeOH); IR (ATR): ν = 3371w, 2934vs, 2865m, 1706m, 1643s, 1524s, 1519s, 1466s, 1450s, 1387m, 1376s, 1356s, 1328m, 1319m, 1301m, 1277m, 1245s, 1197s, 1161m, 1131m, 1109m, 1085m, 1046s, 1035s, 1003m, 987m, 973m; ^{1}H-NMR (500 MHz, $CDCl_3$) δ = 3.67–3.55 (m, 2H, 31-H_a, 31-H_b), 3.27 (td, J = 11.2, 3.9 Hz, 2H, 1-H_a, 1-H_b), 3.17–3.07 (m, 3H, 32-H_a, 32-H_b, 3-H), 2.20–2.12 (m, 1H, 13-H), 2.09 (s, 3H, 28-H), 2.10–2.02 (m, 1H, 16-H_a), 2.00–1.94 (m, 1H, 18-H), 1.91–1.73 (m, 6H, 12-H_a, 34-H, 34′-H, 22-H_a), 1.62–1.06 (m, 22H, 19-H_a, 33-H, 33′-H, 15-H_a, 15-H_b, 2-H_a, 16-H_b, 6-H_a, 22-H_b, 12-H_b, 11-H_a, 6-H_b, 7-H_a, 35-H, 7-H_b, 21-H_a, 9-H, 11-H_b, 21-H_b), 1.05–0.92 (m, 1H, 2-H_b), 0.90 (s, 3H, 27-H), 0.89 (s, 3H, 23-H), 0.82 (s, 3H, 25-H), 0.85–0.78 (m, 1H, 19-H_b), 0.74 (s, 3H, 24-H), 0.68 (s, 3H, 26-H), 0.63–0.58 (m, 1H, 5-H); ^{13}C-NMR (126 MHz, $CDCl_3$) δ = 212.7 (C-20), 177.8 (C-30), 78.7 (C-3), 56.9 (C-32), 55.6 (C-17), 55.3 (C-5), 54.0 (C-33), 51.1 (C-19), 50.4 (C-9), 50.0 (C-18), 42.1 (C-14), 40.6 (C-8), 38.8 (C-4), 38.6 (C-1), 37.6 (C-22), 37.1 (C-10), 36.7 (C-13), 34.3 (C-7), 34.2 (C-31), 32.1 (C-16), 30.0 (C-28), 29.4 (C-21), 28.4 (C-12), 28.0 (C-23), 27.3 (C-15), 27.2 (C-2), 22.8 (C-34), 21.8 (C-35), 20.90 (C-11), 18.2 (C-6), 16.1 (C-24), 16.1 (C-25), 15.4 (C-26), 14.6 (C-27); MS (ESI, MeOH): m/z = 568.5 (100%, $[M + H]^+$); analysis calcd. for $C_{36}H_{60}N_2O_3$ (568.89): C 76.01, H 10.63, N 4.92; found: C 75.93, H 10.46, N 4.71.

5. Conclusions

In this study, 40 triterpenoic acid amides (**11–50**), with different triterpenoic backbones and various amine residues were synthesized and subjected to Ellman's assay to determine their potential as inhibitors of cholinesterases. Furthermore, some enzyme kinetic studies were performed. Thus, systematic variation of the amine substituent led to analogs possessing the same or even better BChE-inhibiting properties as standard galantamine hydrobromide. Outstanding derivatives were 2-pyrrolidin-1-ylethyl substituted compounds **34** (from **BA**), **36** (from **UA**), and **37** (from **OA**), showing K_i values of 0.39 ± 0.04 μM, 0.47 ± 0.08 μM, and 0.55 ± 0.02 μM, respectively. Furthermore, Ellman's assay revealed several platanic acid compounds as excellent inhibitors of BChE. Particularly, **40**, **45**, and **50** were great inhibitors, showing inhibition rates even in the nanomolar range. The most active compound in the test was a hybrid holding a platanic acid backbone and a pyrrolidinyl residue. For (3β)-*N*-(2-pyrrolidin-1-ylethyl)-3-acetyloxy-20-oxo-30- norlupan-28-amide (**35**), inhibition constants $K_i = 0.07 \pm 0.01$ μM and $K_i' = 2.38 \pm 0.48$ μM have been determined. The results obtained in the biological assay can be explained by appropriate molecular modeling calculations. All active compounds were mixed-type BChE inhibitors with a dominating competitive part ($K_i < K_i'$). The best inhibitor for acetylcholinesterase was a betulinic acid derived piperidinyl derivative (49), acting as a mixed-type inhibitor showing $K_i = 1.00 \pm 0.09$ μM and $K_i' = 1.42 \pm 0.08$ μM, respectively.

Supplementary Materials: Supplementary data related to this article can be found online.

Author Contributions: A.L., M.K., and R.C., conceived and designed the experiments; M.K. and I.S. performed the experiments; A.L., performed the biological assays and experiments; W.B., performed the molecular modeling; A.L., M.K., and R.C., analyzed the data and wrote the paper.

Funding: We acknowledge the financial support within the funding program Open Access Publishing by the German Research Foundation (DFG).

Acknowledgments: We'd like to thank R. Kluge for measuring the ESI-MS spectra and D. Ströhl and his team for the NMR spectra. Thanks are also due to J. Wiese and V. Simon for the IR spectra and optical rotations.

Conflicts of Interest: The authors declare no conflict of interest.

References

1. Alzheimer, A. Über eine eigenartige Erkrankung der Hirnrinde. *Allg. Z. Psychiatr. Psych.-Gerichtl. Med.* **1907**, *64*, 146–148.
2. Davies, P. Selective loss of central cholinergic neurons in Alzheimer's disease. *Lancet* **1976**, *308*, 1403. [CrossRef]
3. Peron, G.; Marzaro, G.; Dall'Acqua, S. Known Triterpenes and their Derivatives as Scaffolds for the Development of New Therapeutic Agents for Cancer. *Curr. Med. Chem.* **2018**, *25*, 1259–1269. [CrossRef] [PubMed]
4. Rezanka, T.; Siristova, L.; Sigler, K. Sterols and Triterpenoids with Antiviral Activity. *Anti-Infect. Agents Med. Chem.* **2009**, *8*, 193–210. [CrossRef]
5. Xiao, S.; Tian, Z.; Wang, Y.; Si, L.; Zhang, L.; Zhou, D. Recent progress in the antiviral activity and mechanism study of pentacyclic triterpenoids and their derivatives. *Med. Res. Rev.* **2018**, *38*, 951–976. [CrossRef] [PubMed]
6. Cunha, L.C.S.; e Silva, M.L.A.; Furtado, N.A.J.C.; Vinhólis, A.H.C.; Martins, C.H.G.; da Silva Filho, A.; Cunha, W.R. Antibacterial Activity of Triterpene Acids and Semi-Synthetic Derivatives against Oral Pathogens. *Z. Naturforsch. C* **2007**, *62*, 668–672. [CrossRef]
7. Wolska, K.; Grudniak, A.; Fiecek, B.; Kraczkiewicz-Dowjat, A.; Kurek, A. Antibacterial activity of oleanolic and ursolic acids and their derivatives. *Open Life Sci.* **2010**, *5*, 543–553. [CrossRef]
8. Song, Q.-Y.; Qi, W.-Y.; Li, Z.-M.; Zhao, J.; Chen, J.-J.; Gao, K. Antifungal activities of triterpenoids from the roots of Astilbe myriantha Diels. *Food Chem.* **2011**, *128*, 495–499. [CrossRef] [PubMed]
9. Jamila, N.; Khairuddean, M.; Yeong, K.K.; Osman, H.; Murugaiyah, V. Cholinesterase inhibitory triterpenoids from the bark of *Garcinia hombroniana*. *J. Enzyme Inhib. Med. Chem.* **2015**, *30*, 133–139. [CrossRef] [PubMed]
10. Ahmad, Z.; Mehmood, S.; Ifzal, R.; Malik, A.; Afza, N.; Ashraf, M.; Jahan, E. A New Ursane-type Triterpenoid from Salvia santolinifolia. *Turk. J. Chem.* **2007**, *31*, 495–501.

11. Schwarz, S.; Loesche, A.; Lucas, S.D.; Sommerwerk, S.; Serbian, I.; Siewert, B.; Pianowski, E.; Csuk, R. Converting maslinic acid into an effective inhibitor of acylcholinesterases. *Eur. J. Med. Chem.* **2015**, *103*, 438–445. [CrossRef] [PubMed]
12. Heller, L.; Kahnt, M.; Loesche, A.; Grabandt, P.; Schwarz, S.; Brandt, W.; Csuk, R. Amino derivatives of platanic acid act as selective and potent inhibitors of butyrylcholinesterase. *Eur. J. Med. Chem.* **2017**, *126*, 652–668. [CrossRef] [PubMed]
13. Loesche, A.; Köwitsch, A.; Lucas, S.D.; Al-Halabi, Z.; Sippl, W.; Al-Harrasi, A.; Csuk, R. Ursolic and oleanolic acid derivatives with cholinesterase inhibiting potential. *Bioorganic Chem.* **2019**, *85*, 23–32. [CrossRef] [PubMed]
14. Wiemann, J.; Loesche, A.; Csuk, R. Novel dehydroabietylamine derivatives as potent inhibitors of acetylcholinesterase. *Bioorg. Chem.* **2017**, *74*, 145–157. [CrossRef] [PubMed]
15. Loesche, A.; Wiemann, J.; Rohmer, M.; Brandt, W.; Csuk, R. Novel 12-Hydroxydehydroabietylamine derivatives act as potent and selective butyrylcholinesterase inhibitors. *Bioorg. Chem.*. submitted.
16. Perry, E.K.; Perry, R.H.; Blessed, G.; Tomlinson, B.E. Changes in brain cholinesterases in senile dementia of Alzheimer type. *Neuropathol. Appl. Neurobiol.* **1978**, *4*, 273–277. [CrossRef] [PubMed]
17. Dixon, M. The determination of enzyme inhibitor constants. *Biochem. J.* **1953**, *55*, 170–171. [CrossRef] [PubMed]
18. Cornish-Bowden, A. A simple graphical method for determining the inhibition constants of mixed, uncompetitive and non-competitive inhibitors. *Biochem. J.* **1974**, *137*, 143–144. [CrossRef] [PubMed]
19. Lineweaver, H.; Burk, D. The Determination of Enzyme Dissociation Constants. *J. Am. Chem. Soc.* **1934**, *56*, 658–666. [CrossRef]
20. Sommerwerk, S.; Csuk, R. Convenient and chromatography-free partial syntheses of maslinic acid and augustic acid. *Tetrahedron Lett.* **2014**, *55*, 5156–5158. [CrossRef]
21. Sommerwerk, S.; Heller, L.; Serbian, I.; Csuk, R. Straightforward partial synthesis of four diastereomeric 2,3-dihydroxy-olean-12-en-28-oic acids from oleanolic acid. *Tetrahedron* **2015**, *71*, 8528–8534. [CrossRef]
22. Kahnt, M.; Heller, L.; Grabandt, P.; Al-Harrasi, A.; Csuk, R. Platanic acid: A new scaffold for the synthesis of cytotoxic agents. *Eur. J. Med. Chem.* **2018**, *143*, 259–265. [CrossRef] [PubMed]
23. Kahnt, M.; Al-Harrasi, A.; Csuk, R. Ethylenediamine Derived Carboxamides of Betulinic and Ursolic Acid as Potential Cytotoxic Agents. *Molecules* **2018**, *23*, 2558. [CrossRef] [PubMed]
24. Deng, S.-L.; Baglin, I.; Nour, M.; Cavé, C. Synthesis of phosphonodipeptide conjugates of ursolic acid and their homologs. *Heteroat. Chem.* **2008**, *19*, 55–65. [CrossRef]
25. Wong, M.H.L.; Bryan, H.K.; Copple, I.M.; Jenkins, R.E.; Chiu, P.H.; Bibby, J.; Berry, N.G.; Kitteringham, N.R.; Goldring, C.E.; O'Neill, P.M.; et al. Design and Synthesis of Irreversible Analogues of Bardoxolone Methyl for the Identification of Pharmacologically Relevant Targets and Interaction Sites. *J. Med. Chem.* **2016**, *59*, 2396–2409. [CrossRef] [PubMed]
26. Siewert, B.; Pianowski, E.; Obernauer, A.; Csuk, R. Towards cytotoxic and selective derivatives of maslinic acid. *Bioorg. Med. Chem.* **2014**, *22*, 594–615. [CrossRef] [PubMed]
27. Petrenko, N.I.; Elantseva, N.V.; Petukhova, V.Z.; Shakirov, M.M.; Shul'ts, E.E.; Tolstikov, G.A. Synthesis of betulonic acid derivatives containing amino-acid fragments. *Chem. Nat. Compd.* **2002**, *38*, 331–339. [CrossRef]
28. Vystrčil, A.; Buděšínský, M. Triterpenes. XVI. Unusual epimerisation of the C(19)-acetyl group of 20-oxo-30-norlupane derivatives. *Collect. Czech. Chem. Commun.* **1970**, *35*, 295–311. [CrossRef]
29. Heller, L.; Knorrscheidt, A.; Flemming, F.; Wiemann, J.; Sommerwerk, S.; Pavel, I.Z.; Al-Harrasi, A.; Csuk, R. Synthesis and proapoptotic activity of oleanolic acid derived amides. *Bioorg. Chem.* **2016**, *68*, 137–151. [CrossRef] [PubMed]

Sample Availability: Samples of all compounds are available from the authors.

molecules

MDPI

Article

Identification and Characterization of the Caspase-Mediated Apoptotic Activity of *Teucrium mascatense* and an Isolated Compound in Human Cancer Cells

Neena Gopinathan Panicker [1,†], Sameera Omar Mohammed Saeed Balhamar [1,†], Shaima Akhlaq [1], Mohammed Mansour Qureshi [1], Tania Shamim Rizvi [2], Ahmed Al-Harrasi [2], Javid Hussain [3] and Farah Mustafa [1,*]

1 Department of Biochemistry, College of Medicine & Health Sciences, United Arab Emirates (UAE) University, Al Ain 20000, UAE; ngpanicker@uaeu.ac.ae (N.G.P.); mlt1987@hotmail.com (S.O.M.S.B.); 201570084@uaeu.ac.ae (S.A.); mohammad.qureshi@uaeu.ac.ae (M.M.Q.)

2 Natural and Medical Sciences Research Center, University of Nizwa, Nizwa, Oman; taniarizvi722003@gmail.com (T.S.R.); aharrasi@unizwa.edu.om (A.A.-H.)

3 Department of Biological Sciences & Chemistry, College of Arts and Sciences, University of Nizwa, Nizwa, Oman; javidhej@unizwa.edu.om

* Correspondence: fmustafa@uaeu.ac.ae; Tel.: +971-3-713-7509; Fax: +971-3-767-2033

† These authors contributed equally to this project.

Academic Editors: Eva E. Rufino-Palomares and José Antonio Lupiáñez

check for updates

Received: 16 January 2019; Accepted: 4 March 2019; Published: 11 March 2019

Abstract: Plants of the genus *Teucrium* (Lamiaceae or Labiatae family) are known historically for their medicinal value. Here, we identify and characterize the anticancer potential of *T. mascatense* and its active compound, IM60, in human cancer cells. The anti-proliferative effect of a *T. mascatense* methanol extract and its various fractions were analyzed in MCF-7 and HeLa cells in a dose- and time dependent manner. The dichloromethane fraction (TMDF) was observed to be the most effective with cytotoxicity against a more expanded series of cell lines, including MDA-MB-231. A time and dose-dependent toxicity profile was also observed for IM60; it could induce rapid cell death (within 3 h) in MCF-7 cells. Activation of caspases and PARP, hallmarks of apoptotic cell death pathways, following treatment with TMDF was demonstrated using western blot analysis. Inversion of the phosphatidylserine phospholipid from the inner to the outer membrane was confirmed by annexin V staining that was inhibited by the classical apoptosis inhibitor, Z-VAK-FMK. Changes in cell rounding, shrinkage, and detachment from other cells following treatment with TMDF and IM60 also supported these findings. Finally, the potential of TMDF and IM60 to induce enzymatic activity of caspases was also demonstrated in MCF-7 cells. This study, thus, not only characterizes the anticancer potential of *T. mascatense*, but also identifies a lead terpenoid, IM60, with the potential to activate anticancer cell death pathways in human cancer cells.

Keywords: *Teucrium mascatense*; natural plant products; anticancer activity; breast and cervical cancer; apoptosis; caspases

1. Introduction

Cancer is the second major cause of mortality worldwide, responsible for an estimated 9.6 million deaths in 2018, and accounting for one out of every six deaths [1]. This burden is expected to rise from 18.1 million worldwide to 29.5 million by 2040 [2]. The economic burden of cancer is no less, where only in the US, it was estimated to be nearly $1.16 trillion in 2010 alone [3]. The use of plant-derived compounds for cancer therapy is widely prevalent, with more than 60% of clinically approved anticancer drugs being derivatives of medicinal plants [4,5]. However, surprisingly, only 6% of the nearly 250,000 plant species known to exist have been screened for their biological activity [6]. Based on estimates of the World Health Organization (WHO), around 80% of the world's population depends on herbal medicines, especially natives of Latin America, Asia, and Africa [7], and more than 3000 plant species are currently known to treat cancer [8]. Considering the wealth of untapped knowledge available in the realm of traditional/alternative medicine, there is an urgent need to use advanced screening methods to identify and characterize the anticancer potential of plant extracts [9]. Interestingly, most of the therapeutic drugs currently used in the treatment of breast cancer, the most frequent cause of cancer and cancer-related mortality among women [2], are originally derived from plants [7]. These include paclitaxel and docetaxel (*Taxus baccata*), etoposide (*Podophyllum peltatum*), camptothecin (*Camptotheca acuminata*), and vinblastine and vinorelbine (*Vinca rosea*). In fact, results from recent clinical trials have established docetaxel as the most active single agent in the treatment of advanced metastatic breast cancer, either as first or second-line therapy [10]. Therefore, the skepticism often associated with the translation of compounds derived from plant extracts into commercially viable drugs is unsubstantiated.

Teucrium mascatense (*T. mascatense*) is a medicinal plant of the genus *Teucrium*, belonging to Lamiaceae or Labiatae family. The *Teucrium* genus is comprised of around 300 species distributed over central Europe, Western Asia, the Mediterranean region, North Africa, and the Arabian Peninsula [11–17]. The medicinal value of *Teucrium* species has been known since the times of Socrates and Jalinous, and plants belonging to this genus have been used in both traditional and modern medicine owing to their bioactive constituents [15–19]. Species of *Teucrium* are known to contain tannins, glycosides, phenols, steroids, and terpenoids, with strong biological activities, such as antibacterial, antipyretic, anti-inflammatory, anti-diabetic, anti-spasmodic, analgesic, lipolysis, and antioxidant actions [19–31]. One species of *Teucrium* from the same family, found in areas like Sardinia and Baronia of Siniscola, has been used as an antimalarial agent [32].

Not much is known about the anticancer potential of *T. mascatense*. Most of the anticancer studies on this plant have been done on *Teucrium polium* (*T. polium*), one of the most widely studied plants of the *Teucrium* genus [33]. It has been shown to be an effective and safe chemo-sensitizing agent as it can potentiate the anti-proliferative and apoptotic effects of various chemotherapeutic drugs, including vincristine, vinblastine, and doxorubicin [34]. *T. polium* may also have anticancer potential that can be attributed to the presence of flavonoids and diterpenoids [19,35]. In addition, secondary metabolites present in this species have been shown to have toxic effects against cancer cells [36–38]. Kandouz et al., have shown that extracts of *T. polium* can not only inhibit proliferation of prostate cancer cells, but also inhibit their invasion and motility by altering the expression and localization of E-cadherin and catenins [39]. Finally, a recent study using concentrates of *T. polium* in rats has shown significant anticancer activity against hepatocellular carcinomas [40].

Compared to these, there is a dearth of literature on the anticancer potential of *T. mascatense*. Initial screening of the *T. mascatense* plant extracts showed that it could inhibit growth of human breast cancer cells [41]. Thus, the aim of the current study was to characterize this anticancer potential of *T. mascatense* further in a comprehensive manner. Our work demonstrates that *T. mascatense* can induce apoptotic activity in human breast cells. Furthermore, we go on to demonstrate that an active compound isolated from *T. mascatense* has anti-proliferative and anti-apoptotic activity against breast cancer cells in vitro, leading to the identification of a potential lead compound in the search for natural compounds against cancer.

2. Results

2.1. Crude Methanolic Extract of T. mascatense and Some of Its Fractions Induce Cytotoxic Effects in Human Normal and Cancer Cell Lines

To determine the anticancer potential of *T. mascatense*, its methanol extract (TMME) along with five of its organic/aqueous fractions were analyzed for their effects on cancer cell proliferation, as described in the Materials and Methods section. The different organic/aqueous fractions of the crude methanol extract were tested to determine which solvents were better at isolating the relevant biologically active content of the plant [42]. The initial screening was limited to two types of human cancer cell lines: MCF-7, a breast cancer cell line [43], and HeLa, a cervical carcinoma cell line [44] (Table 1). This allowed us to determine the general potential of these extract/fractions to affect cell proliferation of two different types of human cancer cell lines. The samples were analyzed at four different concentrations (25, 50, 125, 250 μg/mL), at 24 and 72 h post-treatment. After 24 h of treatment, TMME and two of its organic fractions prepared by using dichloromethane and *n*-hexane solvents (TMDF and TMHF), were observed to be effective against MCF-7 cell line by inducing ≥20% inhibition of cell proliferation at a concentration of 250 μg/mL. None of these three showed cytotoxic effects on HeLa cells 24 h post treatment. After 72 h, TMDF, TMME, and TMHF showed ≥20% cytotoxic effects on the MCF-7 cell line at concentrations of 125 and 250 μg/mL, and were also active against the HeLa cell line at a dose of 250 μg/mL (Table 1).

Table 1. Summary of the first screening of *T. mascatense* extract/fractions using the MTT assay.

Extract/Fractions	MCF-7 (24/72 HRS) Breast Cancer Cells (μg/mL) *	HeLa (24/72 HRS) Cervical Cancer Cells (μg/mL) *
TMHF	24HR: 250 72HR: 125/250	24HR: X 72HR: 250
TMDF	24HR: 250 72HR: 125/250	24HR: X 72HR: 250
TMEF	X	X
TMBF	X	X
TMME	24HR: 250 72HR: 125/250	24HR: X 72HR: 250
TMAF	X	X

X = Extract/fractions displaying ≤ 20% inhibition of proliferation. * = Numbers denote extract/fractions concentration (in μg per mL) at which ≥ 20% inhibition of proliferation was observed. *Abbreviations*: TMHF: *Teucrium mascatense n*-hexane fraction; TMDF: *Teucrium mascatense* dichloromethane fraction; TMEF: *Teucrium mascatense* ethyl acetate fraction; TMBF: *Teucrium mascatense n*-butanol fraction; TMME: *Teucrium mascatense* methanol extract; TMAF: *Teucrium mascatense* aqueous fraction.

Based on our experience with the anti-cancer potential of crude methanol extracts and their different fractions from several other plants where the dichloromethane solvent was the most consistent in its ability to induce cytotoxic effects in both MCF 7 and HeLa cell lines, TMDF was chosen for further testing in a more expanded series of cell lines, including MCF-10A and MDA-MB-231. MCF10A is a normal human mammary epithelial cell line [45] and was used to allow comparison of the effect of TMDF on normal versus cancer cell lines, while MDA-MB-231 is a cell line from a triple receptor negative breast cancer tissue [46]. Such breast cancers are much harder to treat due to their inability to respond to therapies directed against hormone receptors [47]; thus, this cell line allowed us to test for natural compounds that may have anti-proliferation activity against them.

The MTT assay was performed on all the cell lines chosen in a dose-dependent manner using three different concentrations of extract/fractions (50, 125, 250 μg/mL) after 24, 48, and 72 h of treatment. Figure 1 shows the dose-dependency of each cell line and the time course of cell death observed after normalizing cell proliferation to the effects of DMSO alone (the solvent used for solubilizing TMDF). As can be seen, TMDF could induce cytotoxic effects in both the breast and cervical cancer cell lines in a statistically significant manner (Figure 1). However, the normal breast epithelial cell line, MCF-10A, was the most sensitive to TMDF, while the three cancer cell lines showed comparable dose response to TMDF (Figure 1C). Calculation of the dose that caused 50% inhibition of proliferation (IC_{50}) confirmed these observations. As can be seen from the table in Figure 1, the IC_{50} value for the 72-h time point for MCF-10A was the least (45.83 μg/mL), followed by that for HeLa (196.4 μg/mL), MCF-7 (227 μg/mL) and MDA-MB-231 cells (232.8 μg/mL). On the other hand, time course analysis of cell viability revealed that the cytotoxicity profile for the normal MCF-10A cells was comparable to the other two breast cancer cell lines, MCF-7 and MDA-MB-231, while the HeLa cells were the slowest in response to TMDF-induced cell death (Figure 1F).

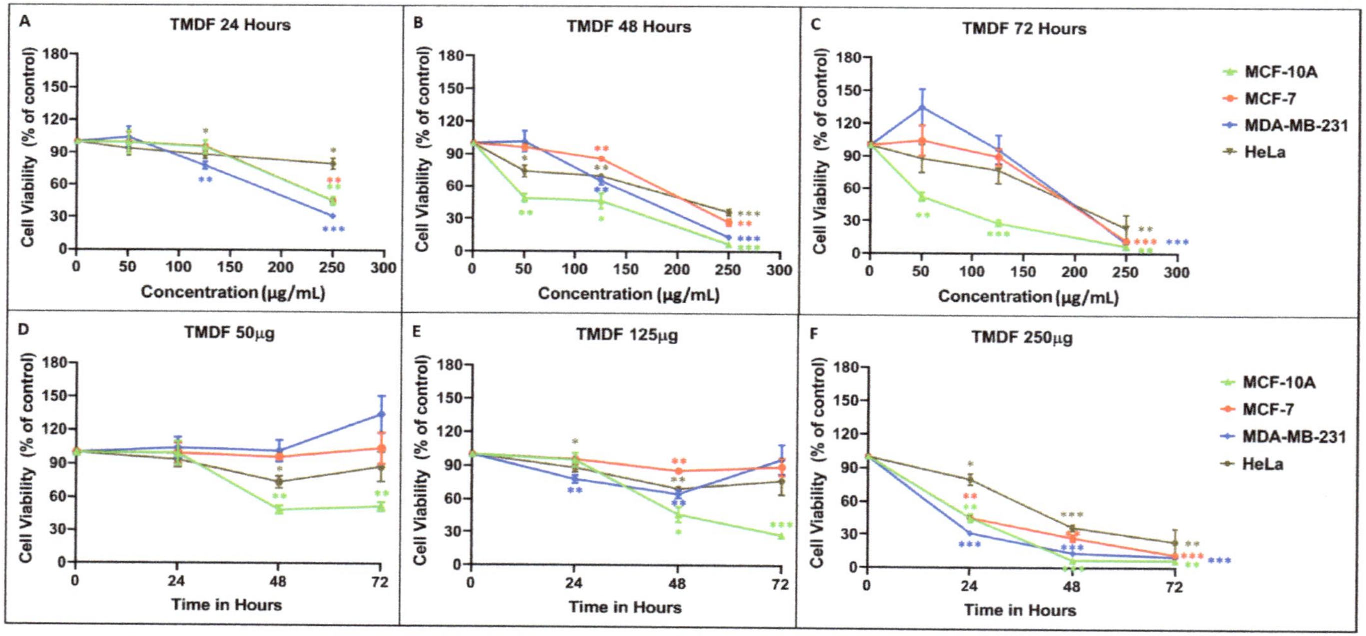

	MCF-10A	MCF-7	MDA-MB-231	Hela
IC_{50} (μg/mL)	45.83	227.00	232.80	196.40

Figure 1. Cytotoxic effects of different concentration of *T. mascatense* dichloromethane fraction (TMDF) using MTT assay on MCF-10A, MCF-7, MDA-MB-231, and HeLa cells after (**A**,**D**) 24, (**B**,**E**) 48, and (**C**,**F**) 72 h post treatment. * indicates statistically significant differences between the DMSO- and TMDF-treated samples (* $p < 0.05$; ** $p < 0.01$ but > 0.001; *** $p < 0.001$). The IC_{50} values (in μg/mL) for the 72-h time points are shown in the table below the figure.

To eliminate the possibility of the observed cytotoxicity being due to the organic solvents used in the extraction process, MTT assay was also conducted on the three cancer cell lines (MCF-7, MDA-MB-231, and HeLa) using only the individual solvents starting at 1:500 dilution. None of the organic solvents showed any cytotoxicity on the tested cell lines (data not shown). This was further validated by nuclear magnetic resonance (NMR) analysis of the extract and fractions which detected no residual organic solvents in these samples (data not shown). These observations, along with the fact that only three of the five organic extract/fractions showed cytotoxicity on the tested cell lines, (Table 1) confirms that the detected anti-proliferative effects on the cell lines were not due to any lingering solvents that may have been left in the extract/fractions; rather, it was due to the compounds present in the extract and fractions themselves.

2.2. *TMDF Activates Key Apoptotic Proteins in Breast Cancer Cells*

Next, we determined whether the cytotoxicity being observed in the cancer cell lines was due to apoptosis, the primary cell death pathway activated by anticancer compounds [7,48]. The activation of various caspases and downstream effector protein PARP (poly ADB ribose polymerase) are hallmarks of apoptosis [49]. Thus, the classical technique of western blot analysis was used to detect activation of these proteins involved in apoptosis 24 h post treatment with different concentrations of TMDF. As shown in Figure 2A, the expression and cleavage of caspase 7 and PARP proteins was successfully detected, demonstrating that TMDF had the potential to activate a caspase-dependent mechanism of inducing apoptosis in MCF-7 cells. Some degradation of actin could be observed at the 250 μg/mL concentration, suggesting late stages of apoptosis when actin degradation can be expected [50]. Activation of caspase 8 and 9 could also be observed by the steady disappearance of the procaspase 8 and 9 bands; however, the actual cleaved products of these caspases could not be detected (Figure 2A). This is most likely due to the well-known labile nature of the activated caspase cleavage products that requires correct timing and concentration to be visible and the poorer specificity of the antibodies for the cleaved products [51].

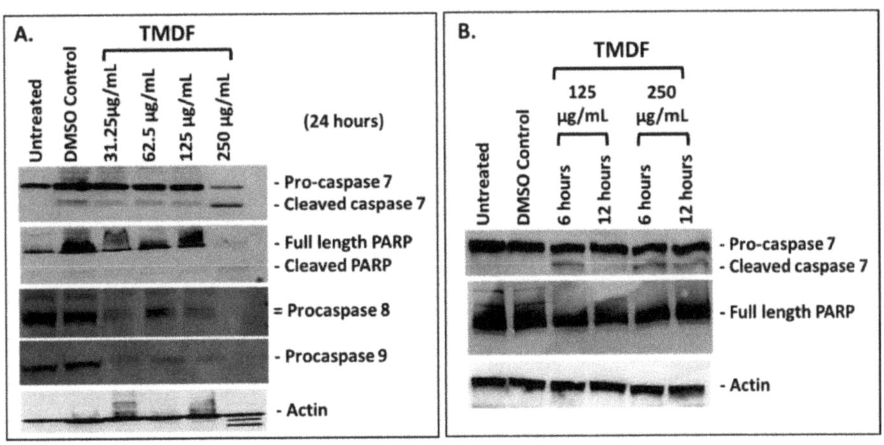

Figure 2. Western blot analysis of MCF-7 cells treated with the dichloromethane fraction of *T. mascatense* (TMDF): (**A**) for 24 h, (**B**) for 6 and 12 h at the indicated concentrations followed by detection of the specified caspases or poly ADB ribose polymerase (PARP) proteins. Actin antibody was used as a loading control. The "DMSO control" sample contains the same amount of DMSO as present in the 250 μg/mL sample of TMDF.

To ensure that the caspase cleavage being observed was reproducible and not a reflection of generalized cytotoxicity from TMDF, the westerns for caspase 7 and PARP were repeated at earlier time points (6 and 12 h) on cell lysates generated from MCF-7 cells treated with 125 and 250 μg/mL of TMDF. As can be seen in Figure 2B, caspase 7 cleavage could be detected under these conditions where actin was not degraded, confirming that the caspase 7 cleavage observed earlier was bona fide and not due to a generalized cytotoxicity induced by TMDF. Interestingly, under these conditions, PARP cleavage was not visible, most likely due to the early time points studied since PARP cleavage is a later event in apoptosis [49].

2.3. TMDF Can Induce Apoptosis in MCF-7 Cells

Another hallmark of apoptosis is inversion of the inner leaflet of the plasma membrane, exposing the proteins and phospholipids that reside on the inside of the plasma membrane phospholipid bilayer without compromising membrane integrity [49]. This is considered an early event observed in cells undergoing apoptosis in which the cell viability is maintained despite the inversion of the plasma membrane [49]. Therefore, we tested for the ability of TMDF to induce this classical effect on treated cells using the well-established Annexin V/propidium iodide (PI) staining assay and flow cytometry [49]. As can be seen from Figure 3, compared to the untreated cells, DMSO treatment did not affect the viability of the cells (Figure 3A,B). This is in contrast to our positive control (*Boswellia sacra* essential oil) that has a potent ability to induce apoptosis in a rapid manner (Figure 3C) [52]. Treatment of MCF-7 cells with *Boswellia sacra* essential oil led to shift of the cells into the Annexin V-positive, but PI negative quadrants (19.6%), revealing the induction of early apoptotic events (Figure 3C) as well as late apoptosis (28.9%) where the Annexin V population was PI positive (Figure 3H). In comparison, treatment of MCF-7 cells for six hours with 125 μg/mL of TMDF led to 21.85% of the cells to move into the early apoptosis phase (Figure 3D), while treatment with 250 μg/mL lead to a further increase in apoptosis induction with 62.6% of the cells moving into early apoptosis phase and 2.1% into late apoptosis (Figure 3E,H). This dose-dependent increase in cells entering the early apoptotic stage reveals that TMDF has the capability to induce apoptosis in MCF-7 cells.

To confirm whether this was indeed apoptosis, we further tested the ability of the classical apoptosis inhibitor, Z-VAK-FMK to inhibit this process. As can be seen, pretreatment of the MCF-7 cells with 20 μM Z-VAK-FMK led to a significant inhibition of apoptosis in both 125 and 250 μg/mL TMDF-treated cells (Figure 3F,G) with only 1.9 and 9.5% of the cells shifting into the early apoptotic phase compared to 21.85 and 62.6% of the cells without the inhibitor (Figure 3H). These results conclusively show that TMDF has the ability to induce apoptosis in breast cancer cells.

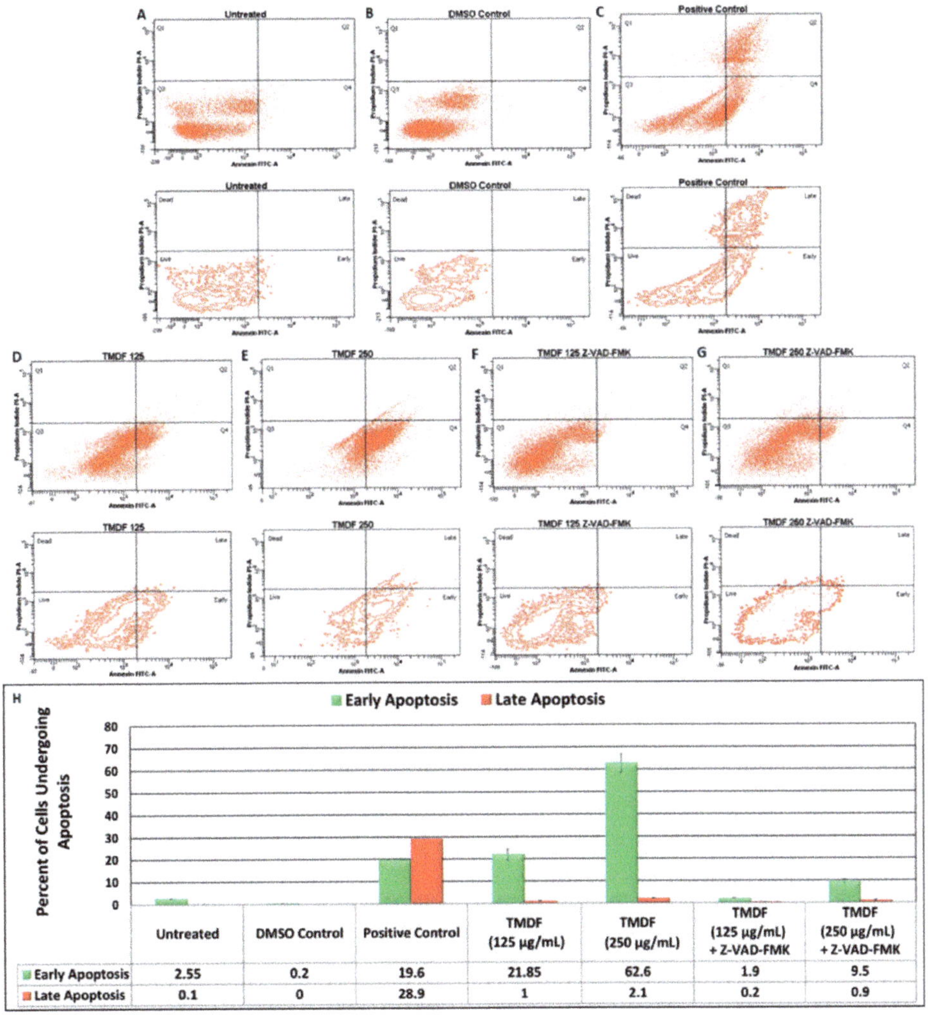

Figure 3. Flow cytometric analysis of Annexin V and propidium iodide-stained MCF-7 cells treated with dichloromethane fraction of *T. mascatense* (TMDF). MCF-7 cells were treated with 125 and 250 µg/mL of TMDF for six hours. (**A**) Untreated cells; (**B**) DMSO control; (**C**) Cells treated with *Boswellia sacra*, a known inducer of apoptosis (positive control) [52]; (**D**) Cells treated with 125 µg/mL of TMDF; (**E**) Cells treated with 250 µg/mL of TMDF; (**F**) Cells pretreated with 20 µM of Z-VAD-FMK followed by six hour treatment with 125 µg/mL of TMDF; (**G**) Cells pre-treated with 20 µM of Z-VAD-FMK followed by six hour treatment with 250 µg/mL of TMDF. (**H**) Column graph that plots the percentage of cells that shift into the early (green bars) and late (red bars) of apoptosis upon the various treatments mentioned above. The "DMSO control" sample contains the same amount of DMSO as present in the 250 µg/mL sample of TMDF.

2.4. Effects of TMDF on the Morphological Properties of Treated Cells

Since the western blot and Annexin V/PI analyses of cells treated with TMDF suggested induction of apoptosis, we next analyzed the effect of TMDF treatment on the morphological characteristics of MCF-7 cells. Such changes started to appear within one day of extract treatment. Figure 4 shows the effect of TMDF on MCF-7 cells 24 h post treatment with 125 and 250 µg/mL of TMDF (Figure 4C,D).

As can be seen, differences were observed in cell morphology of the treated MCF-7 cells, including appearance of cell rounding, cell shrinkage, and detachment of cells from other cells and the plate. Importantly, these changes were absent in untreated (Figure 4A) or DMSO-treated cells (Figure 4B).

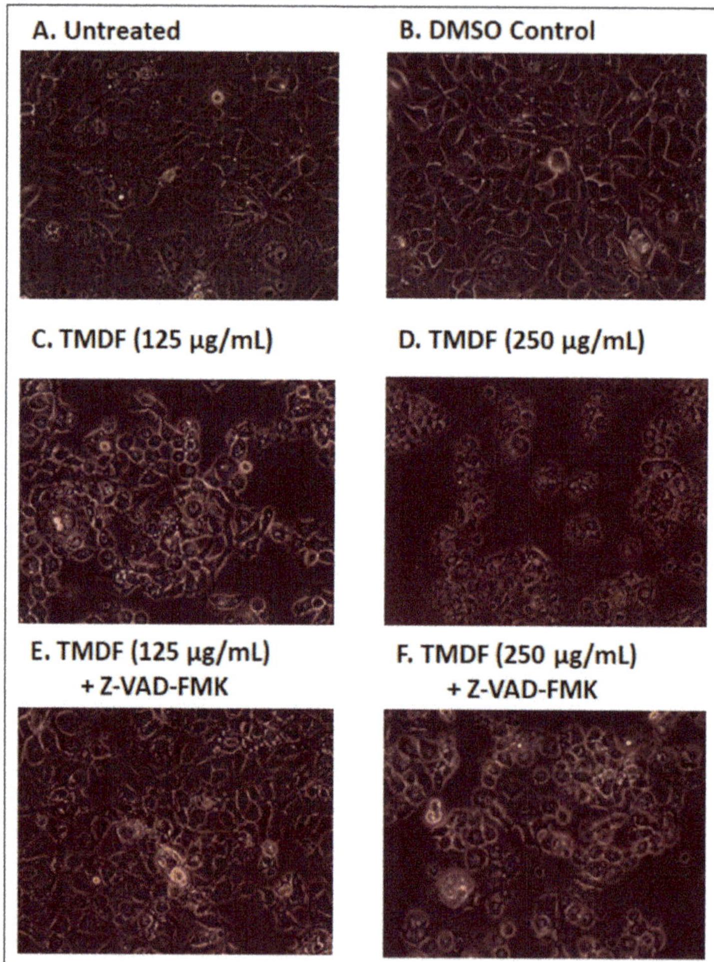

Figure 4. Photomicrographs of MCF-7 cells 24 h after treatment with: (**A**) media alone; (**B**) DMSO; (**C**) 125 μg/mL, and (**D**) 250 μg/mL *T. mascatense* dichloromethane fraction (TMDF). Panels **E** and **F** show the results after a 2-h pretreatement of the cells with 20 μM of the apoptosis inhibitor Z-VAD-FMK before being treated with either (**E**) 125 μg/mL or (**F**) 250 μg/mL of TMDF for 24 h. Magnification: 400×. The "DMSO control" sample contains the same amount of DMSO as present in the 250 μg/mL sample of TMDF.

Next, MCF-7 cells were pretreated with Z-VAD-FMK prior to exposure to TMDF to determine whether the morphological changes associated with apoptosis could be reduced or eliminated upon the inhibitor treatment. As can be seen in panels E and F of Figure 4, pretreatment of the TMDF-treated cultures with Z-VAD-FMK essentially eliminated (Figure 4E) or reduced (Figure 4F) the appearance of the apoptosis-specific morphological changes in cultures treated with 125 and 250 μg/mL of TMDF.

These data support our conclusion that TMDF can activate apoptosis and further suggest that apoptosis is probably the major mechanism of cell death in cultures treated by TMDF.

2.5. TMDF Activates Caspase Activity in Human Breast Cancer Cells

To confirm the potential of TMDF to activate caspases, the Promega caspase GLO assay was used to detect the presence of their enzymatic activity. Towards this end, MCF-7 cells were treated with 250 µg/mL TMDF and tested for the induction of caspase 3/7, 8, and 9 enzymes. Figure 5 shows two independent experiments testing the ability of TMDF to inhibit cell proliferation in MCF-7 cells (blue bars) in parallel with its ability to induce caspase activity. In both experiments, a low cell viability (blue bars) was associated with the induction of high levels of caspase 8 (green bars) and 9 (purple bars) as well as caspase 3/7 (maroon bars) in a statistically significant manner.

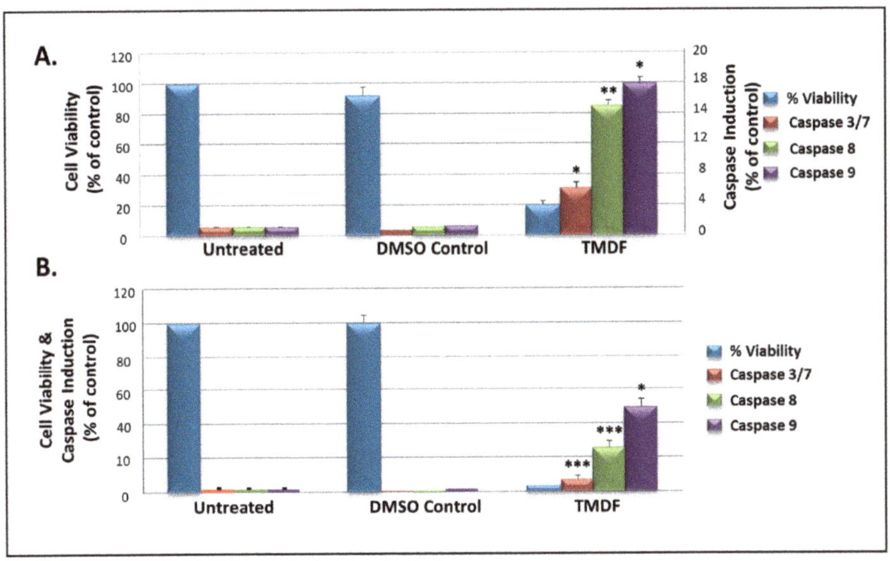

Figure 5. Viability and caspase activity in MCF-7 cells treated with 250 µg/mL of *T. mascatense* dichloromethane fraction (TMDF) for 48 h, as shown in two independent experiments (panels **A** and **B**). * indicates statistically significant differences between the control and treated samples (* $p < 0.05$; ** $p < 0.01$ but > 0.001; *** $p < 0.001$). The "DMSO control" sample contains the same amount of DMSO as present in the 250 µg/mL sample of TMDF.

Since MCF-7 cells do not express any caspase 3 [53], any induction of caspase 3/7 in this system is most likely due to the induction of caspase 7. Results obtained for the induction of caspases were normalized against the number of viable cells in culture (obtained using the Cell Titer-GLO luminescent cell viability assay) to ensure that the enzymatic activity being measured was taking into consideration the ensuing cell death being observed (Figure 5). These observations correlated well with the MTT results presented earlier (Figure 1). Overall, these results strengthen our conclusion that TMDF contains bioactive molecules that can cause cell death via caspase-dependent apoptosis.

2.6. Test of the Anti-Proliferation Effect of IM60 on Breast Cancer Cells

Next, we wanted to study whether a lead compound could be identified from *T. mascatense* with potential anticancer activity. Medicinal plants and their metabolites have an important role in cancer treatment [47,48,54] and their natural or synthetic derivatives could play crucial roles in preventing, slowing, or reversing cancer development [55,56]. Therefore, we screened one purified compound isolated from the dichloromethane fraction of the crude methanolic extract of *T. mascatense*, IM60, to determine its potential anticancer activity [57].

Figure 6 show the results of the dose-dependent effect of IM60 on MCF-7 cells using the MTT assay. The compound was observed to be effective with more than 90% of the MCF-7 cells being killed after 24 h of treatment at a concentration starting at 425 μM (Figure 6). Some activation of cell proliferation was observed at lower concentrations in the 24-h test; however, it was not observed in the 6-h test, suggesting that it could be an artefact. The IC_{50} value was calculated to be 403 μM for the 24-h time point. The earliest differences in cell morphology could be observed at three hours post treatment, eventually leading to complete cell death (Figure 7). Cell shrinkage and rounding and detachment from other cells was noted early on with a loss of cell numbers, which again suggested apoptosis as a possible mechanism of cell death induced by IM60 (Figure 7).

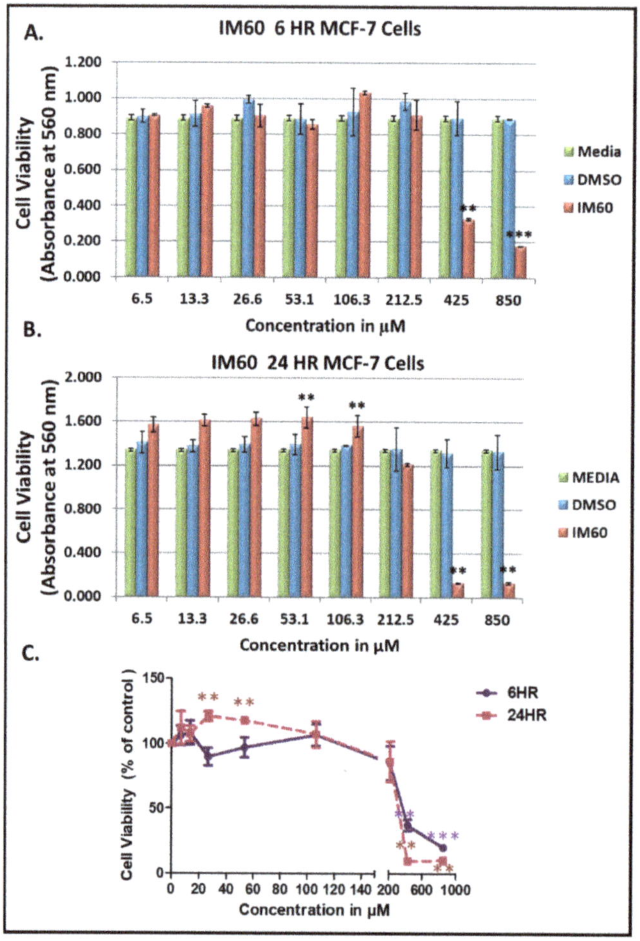

Figure 6. Test of IM60 using MTT assay on MCF-7 cells: (**A**) 6 h post treatment, and (**B**) 24 h post treatment. (**C**) Dose-dependent effect of IM60 on cell viability as a percentage of DMSO-treated cells. * indicates statistically significant differences between the DMSO control and IM60 treated samples (** $p < 0.01$ but > 0.001; *** $p < 0.001$). IM60 was tested twice in the MTT assay in triplicates.

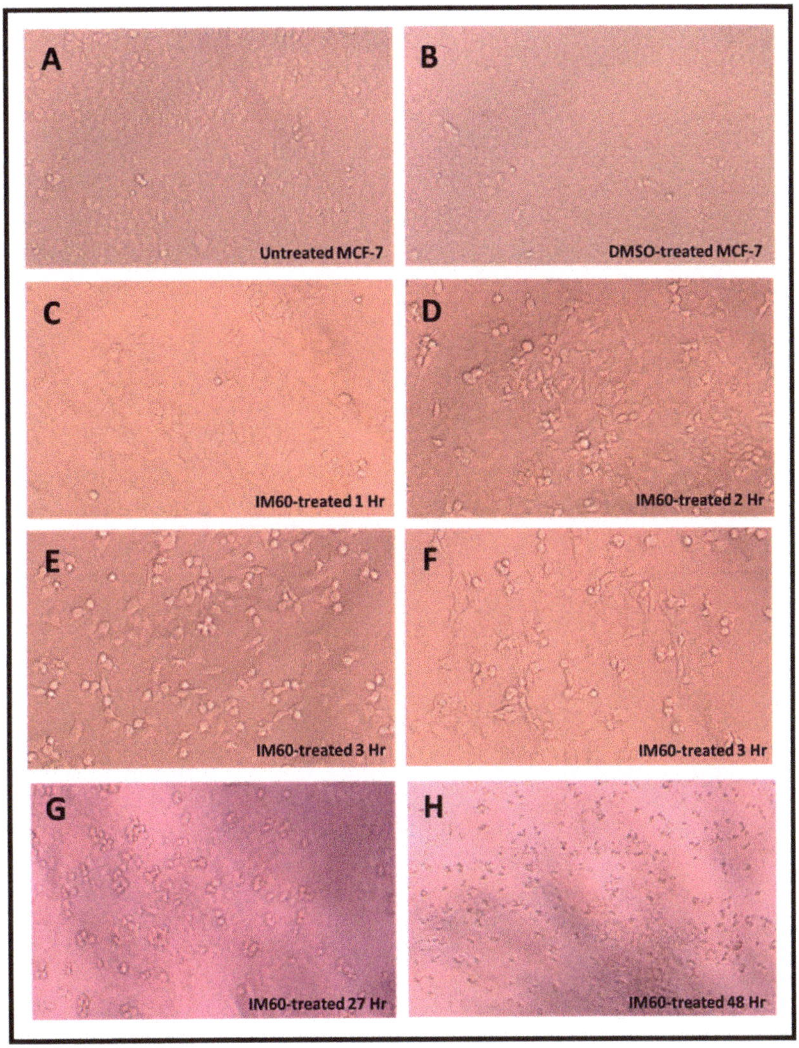

Figure 7. Photomicrographs of MCF-7 cells treated with IM60 compound at 0.425 mM along with its DMSO control for various time points post treatment (magnification 400×). Panels **A–F** are from one experiment, while Panels **G** and **H** are from another experiment that was conducted for a longer time period.

2.7. *IM60 Can Induce Caspase Activity in MCF-7 Cells*

To determine whether IM60 had the potential to induce functional caspase activity, it was tested in the caspase GLO assay conducted on MCF-7 cells at the effective concentration of 425 μM for 6 h. As expected, in comparison to DMSO-treated cells, cell viability was reduced by more than 90% in IM60-treated cells ($p < 0.002$), confirming its cytotoxic effect on MCF-7 (Figure 8). In addition, a statistically significant activation of the effector caspase 7 was noted ($p < 0.03$); however, no significant activation was observed of the initiator caspases 8 or 9. These data suggest that IM60 can cause inhibition of MCF-7 cell proliferation and may have the potential to activate caspase activity as well (Figure 8). Despite the high concentration needed for observing these effects, these are encouraging results since test of a compound and its several derivatives from another medicinal plant tested in parallel did not result in cytotoxicity, and in fact resulted in activation of cancer cell proliferation (personal observations).

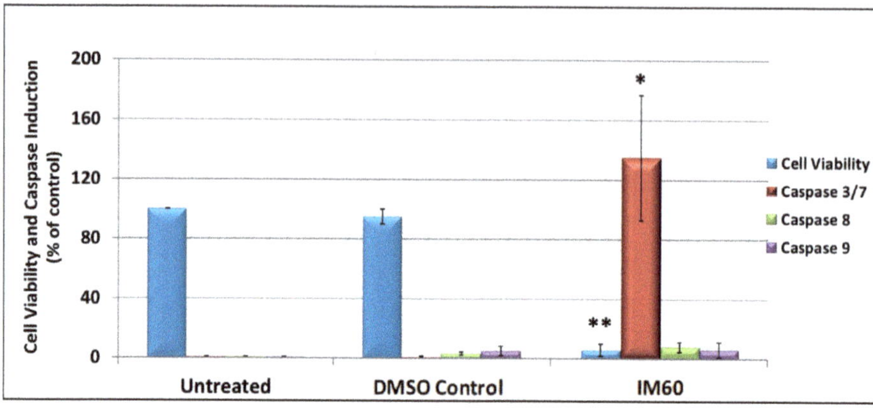

Figure 8. MCF-7 cells were treated with 425 μM of IM60 for 6 h and then assayed for cell viability and induction of different caspase enzyme activities (3/7, 8, and 9). * indicates statistically significant differences between the DMSO-treated and IM60-treated samples (* $p < 0.05$ but > 0.01; ** $p < 0.01$). The "DMSO control" contains the same amount of DMSO as present in the IM60-treated sample.

3. Discussion

This study assessed the anticancer potential of *T. mascatense* by analyzing its different organic and aqueous extract/fractions, including *n*-hexane, dichloromethane, ethyl acetate, *n*-butanol, methanol, and water. Our results revealed that only three of the organic extract/fractions tested (*n*-hexane, methanol, and dichloromethane) had anti-proliferation activity against several human cell lines, including MCF-7, HeLa, and/or MDA-MB-231 (Table 1 and Figure 1), confirming the earlier preliminary test of these extract/fractions [41]. However, we did not find any inhibitory effect of the aqueous extract on either MCF-7 or HeLa cell proliferation, while this fraction showed moderate cytotoxic activity against MDA-MB-231 in the previous study [41]. This difference could be due to the cell line tested. The dichloromethane fraction (TMDF), on the other hand, showed cytotoxic potential against all three cancer cell lines tested, two breast cancer and one cervical cancer (Table 1 and Figure 1). The mechanism of cell death used by TMDF was demonstrated to be apoptosis which was caspase-dependent (Figures 2–5). Furthermore, test of the cytotoxicity of an active constituent of TMDF, IM60, revealed a time and dose-dependent effect (Figure 6); it could induce rapid cell death (within 3 h) with morphological changes reminiscent of apoptosis (Figure 7), and activate caspase 7 enzymatic activity (Figure 8).

Thus, this study adds *T. mascatense* as another *Teucrium* species with anticancer potential. In particular, one study has compared the anti-proliferative effects of six common *Teucrium* species on cell viability of colon cancer cells HTC-116, as well as *T. polium* [58]. Methanolic extracts of all species tested (*T. chamaedrys, T. montanum, T. arduini, T. scordium subsp. Scordium, T. scordium subsp. Scordioides, T. polium* and *T. botrys*) inhibited the proliferation of HTC-116 in MTT assays. Interestingly, the IC_{50} values measured at 72 h post treatment fell between 59–253 μg/mL for all except *T. arduini* which is comparable to our results. Using a crude method to detect apoptosis, their data further suggests that all the tested species could induce apoptosis in these cells. Two other studies have recently tested other species of *Teucrium* for anti-proliferative activity for different types of cancer cells. *T. persicum* can potentially induce apoptosis of prostate cancer cells with inhibitory effects on viability of breast and colon cancer cells [59]. It has also been shown to effect cell migration and epithelial cell morphology in the same study. *T. pruinosum*, on the other hand, has been shown to affect the viability of cervical cancer cell line HeLa [60]. These data suggest that the *Teucrium* genus is a rich source of natural anticancer compounds for further investigation.

While the studies mentioned above have tested the effects of extracts from various *Teucrium* species on cancer cell proliferation, their effect on normal cells has not been studied. The cytotoxic effect of TMDF, on the other hand, has been studied on normal cells and was observed to affect the viability of the normal MCF-10A cells also (Figure 1). This is not very surprising since crude extracts are composed of a number of active biomolecules with different potentials to affect cell proliferation in a cell-specific manner. Thus, it is possible that the biomolecules (or their specific combination) responsible for the cytotoxicity for normal and cancer cells differ, potentially belonging to different compounds that can be separated from each other. That is why the pure compound IM60 was tested to determine if it could affect proliferation of cancer cells. Unfortunately, due to its limited availability, we could not test it on normal cells. In future we plan to test this compound more extensively in both normal and cancer cells to characterize its anti-proliferation potential and determine whether its cytotoxicity can be enhanced for cancer cells and modulated differentially in normal and cancer cells via biochemical derivatization.

The various fractions of TMME tested in this study were made in a sequential manner to separate the active constituents of *T. mascatense* selectively, based on polarity [41,57]. It is well known that the choice of the solvent affects the rate of extraction, the type and quantity of phytochemicals extracted, ease of handling of the extracts, and the health hazards associated with the extraction process [42,61,62]. Among the five tested solvents, dichloromethane [63] and *n*-hexane were observed to be the most effective in isolating the active compounds responsible for the anti-proliferation activity observed in the cancer cells. The remaining organic extracts, and especially the aqueous content, were not cytotoxic, indicating that the biochemical entities responsible for anti-proliferation were primarily hydrophobic in nature. Numerous studies have shown chloroform (a trichloromethane) to be effective in extracting anticancer agents from several medicinal plants, such as *Angelica archangelica*, *Nepeta deflersiana*, and *Solanum nigrum* [64,65].

IM60, tested for its anticancer activity in this study, is a sesquiterpene, 1-isopropanol-4a-methyl-8-methylenedecahydronaphthalene, with the structure shown in Figure 9. It was isolated from the dichloromethane fraction of the crude methanolic extract of *T. mascatense* (TMDF) [57]. Sesquiterpenes belong to a large and diverse group of plant-derived bioactive compounds with promising anticancer, anti-inflammatory, antifungal, antibacterial, and immunosuppressive activities [61,66,67]. This is due to the presence of the decalin ring that imparts great structural and functional variety to these compounds [68] (Figure 9). In one study on the anticancer potential of four different species of *Teucrium*, Menichini et al., identified *T. polium* as the best among the four due to the sesquiterpene content of its essential oil which included compounds such as spathulenol, D-cadinene, caryophyllene, etc. [69]. Identification of a novel sesquiterpene with the ability to activate caspase 7 in this study thus adds to the arsenal of new biomolecules being discovered in the fight against cancer.

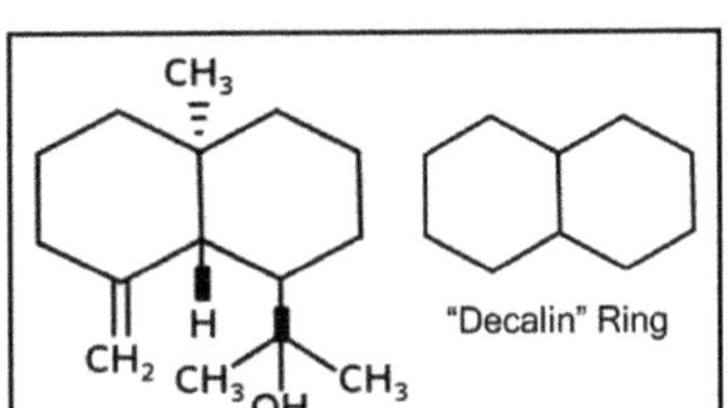

Figure 9. Structure of IM60 with the decalin ring highlighted on the side.

This study shows that caspase-dependent apoptosis is the mechanism of cell death induced by the dichloromethane fraction of *T. mascatense* and perhaps its active component IM60. Apoptosis as a mechanism of cell death has been observed in other plant-derived anticancer compounds as well [7,48]. Most of them induce cell death that may be intrinsic or extrinsic, and caspase and/or p53-dependent or independent. Levitsky and Dembitsky recently analyzed the effect of a large number of plant extracts on breast cancer cells, and observed apoptosis to be one of the most common mechanisms of inducing cell death [7]. For example, genistein was found to induce apoptosis in MCF 7 and T47D breast cancer cell lines. Similarly, oleuropein aglycone found in extra virgin olive oil increased apoptotic cell death by factors of 1.5, 2.5, and 4 in MCF-5, MCF 7/HER, and SK-Br3 cells, respectively. Even "anticancer diets" and crude preparations from a large variety of vegetables have shown to induce apoptosis in breast cancer cell lines. In a study performed on MCF-7 and MDA-MB-231, the cells were pretreated for 72 h with increasing concentrations of *Brassica olearacea* juice, and results suggested that presence of active compounds from cabbage juices activated both apoptosis and necrotic pathway in the breast cancer cells. Similarly, pro-apoptotic effects of green tea extracts and tea catechins have also been reported in tumor cells, both in vitro and in vivo [7,48].

Our preliminary data suggests that TMDF may have the potential to induce autophagy (data not shown), an important cell survival process that is being implicated as a mechanism that prevents neoplastic transformation of cells as well [70,71]. Due to the crude nature of the extracts/fractions tested in this study, it remains to be determined what other cell death pathways can be induced by TMDF and other extract/fractions of *T. mascatense* to understand the full anticancer potential of this plant species.

In this study, we used a cancer cell line model system to study the anticancer potential of our extracts/fractions. Although breast cancer cell lines, in general, are considered to be crude models of the disease as they may not be able to capture the intra- and inter- tumor heterogeneities [72,73], the fact that MCF-7 and MDA-MB-231 cell lines that were used in our analysis, along with T47D, account for more than two-thirds of cell lines used in preclinical studies analyzing breast cancer drugs [73], renders our analysis to be state-of-the-art.

Large-scale screening of plant extracts for anticancer potential has mostly not been very encouraging. In a study of more than 1000 aqueous and organic extracts from 351 species of Brazilian rain forests, only 11 extracts showed any cytotoxicity in MCF-7 cells at a dose of 0.1 mg/mL [74]. An 8-year study of 7500 South African plant extracts found a total of only 50 active extracts when screened against 60 cell lines; none of them was found to be potent in MCF-7, with only 20 presenting moderate inhibiting activity [75]. One reason why they may have missed detecting such activities could be the timing and dosage of the screening. Despite such discouraging data from large-scale analysis, there is reason to believe that concerted, meticulous efforts in the testing of crude plant extracts in a dose and time-dependent manner can lead to identification of active compounds and reveal cellular mechanisms involved in their anticancer potential, as observed in our study.

4. Materials and Methods

4.1. Extraction and Isolation of Plant Material

T. mascatense Boiss (Lamiaceae) was collected in June 2013 from the mountains of *Al-Jabel Al-Akhdar*, Oman. The collection, identification, and extract preparation of the plant has been described previously [41]. A voucher specimen was deposited with the Herbarium of the University of Nizwa, Oman. Briefly, the crude methanolic extract of the whole plant of *T. mascatense* (TMME) was partitioned into five different fractions by solvent-solvent partition with increasing polarity, starting with *n*-hexane (TMHF), dichloromethane (TMDF), a more stable and safer substitute of chloroform with less genotoxic effects [63], ethyl acetate (TMEF), *n*-butane (TMBF), followed by water (TMAF) [41] (Figure 10). IM60 was isolated from the dichloromethane fraction of the crude methanolic extract of *T. mascatense* [57] (Figure 10).

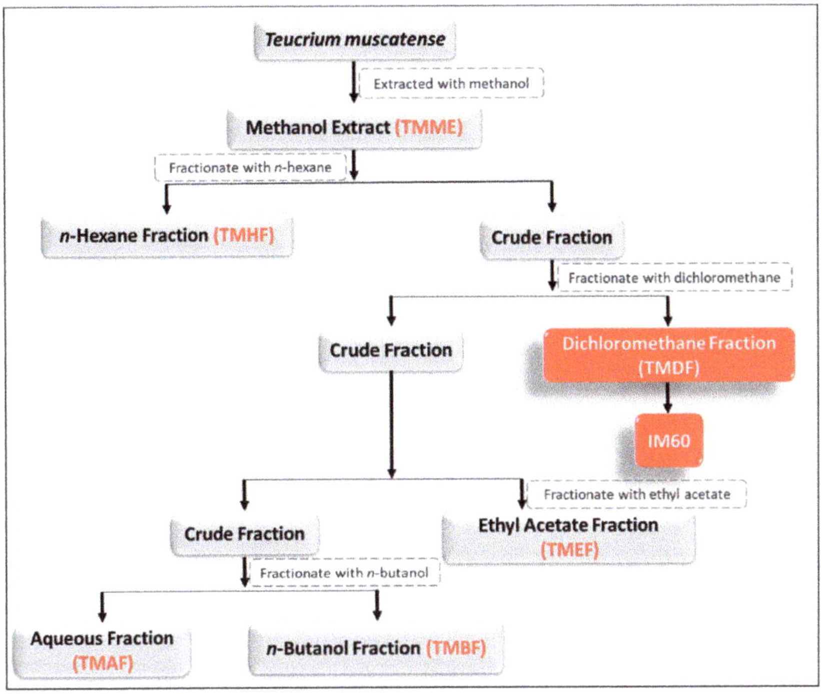

Figure 10. Schematic illustration of the extraction strategy used and isolation of the various fractions and IM60 from the methanolic extract. TMDF and IM60 are highlighted in red as they were the ones characterized in this study in more depth.

To study the anti-proliferative potential of the various extract/fractions of *T. mascatense*, TMME and four of its fractions, TMHF, TMDF, TMEF, TMBF were solubilized in dimethyl sulfoxide (DMSO), while TMAF was solubilized in water to prepare stock solutions at 50 mg/mL (some at lower concentrations due to solubility issues). The stock solutions were stored at −20 °C till the proliferation assays were performed in different cell lines. A compound from *T. mascatense*, IM60, was isolated [57] and tested for its effect on cell proliferation by dissolving it in DMSO. Using the stock solutions of TMME and its fractions, different dilutions were prepared at twice the final concentration (50, 100, 250, and 500 μg/mL), depending upon the cell line used. A final volume of 100 μL of extract/fractions was added to cells plated in 100 μL media on the day of treatment.

4.2. Cell Lines and Culture Medium

Human breast cancer cell lines, MCF-7 (an estrogen and progesterone receptor positive, hormone responsive cell line [43]) and MDA-MB-231 (a triple receptor negative cell line [46]) were cultured in Dulbecco's Modified Eagles Medium (DMEM)/high glucose medium, supplemented with 10% fetal bovine serum (FBS), 10,000 units/mL penicillin/streptomycin (Pen/Strep), and 50 μg/mL gentamicin, while the human cervical cancer cell line, HeLa [44] was grown in DMEM/high glucose medium supplemented with 7% fetal calf serum (FCS), 10,000 units/mL Pen/Strep, and gentamicin (all reagents by HyClone Laboratories, Inc., Logan, UT, USA). The normal breast cancer epithelial cell line, MCF-10A [45] was cultured in DMEM/F12 medium (DMEM/Hams nutrient mixture) and 5% horse serum supplemented with 10,000 units/mL of Pen/Strep, 10 μg/mL insulin, 20 ng/mL epidermal growth factor (EGF), 0.5 mg/mL hydrocortisone, and 100 ng/mL cholera toxin (all reagents by Sigma-Aldrich, St. Louis, MO, USA). The cell lines were maintained at 37 °C in a 5% CO_2 humidified incubator.

4.3. MTT Cell Proliferation Assay

The cytotoxic effect of the extract/fractions was determined using the [3-(4,5-dimethylthiazol-2-yl) -2-5-diphenyltetrazolium bromide] (MTT) colorimetric method [76]. A dose- and time-dependent analysis of the MTT assay was carried out for all the extract/fractions on specified cell lines. Briefly, cells were cultured in 96-well plates at a density of 5000 cells/well/100 μL media. The cell number was determined empirically after testing the proliferative capacity of MCF-7 and HeLa cells in MTT assays using increasing numbers of cells. Using this data, a cell number range was used which provided a linear signal for either MCF-7 or HeLa cell lines.

For the MTT assay, 24 h after plating, the cells were treated with 100 μL of different concentrations of *T. mascatense* extract/fractions dissolved in DMSO, DMSO alone diluted in media to the same concentration as the extract/fractions (DMSO control), or culture media alone for various time points. To measure proliferation in each well, 25 μL of an MTT stock solution (5 mg/mL) was added, and the plates were incubated for 3–4 h in an incubator (at 37 °C). This was followed by decanting the culture media from the plates, dissolving the formazan crystals formed with 200 μL of DMSO, and measuring the absorbance at 560 nm using a plate reader. Cell viability was measured as the percentage of cells treated with DMSO alone, using same concentrations of DMSO that were used to dissolve the extract/fractions (0.2–1%). The IC_{50} values for all the cell lines were calculated using the non-linear regression method of GraphPad Prism 7.4 software (GraphPad Software, San Diego, CA, USA) for the 72-h DMSO-normalized time points or the 24-h time point for IM60.

4.4. Cell Titer-GLO Cell Viability and Caspase-GLO 3/7, 8 & 9 Luminescent Assays

MCF-7 cells were seeded at a density of 5000 cells/well/100 μL in opaque white 96-well plates for tissue culture (Thermo Fisher Scientific, Waltham, MA, USA). After 24 h, cell culture was treated in triplicates with different extract/fractions concentrations or IM60 (in a 50 μL volume) at 37 °C. Cell viability was measured after 48 h using the Cell Titer-GLO Luminescent Cell Viability Assay or the Promega Caspase 3/7, 8 & 9 GLO Assays (Promega Corporation, Fitchburg, WI, USA), according to the manufacturer's directions. Luminescence was measured using the Infinite M200 Pro Tecan plate reader. Data were plotted as percent cell viability of treated groups compared to DMSO-treated cells, the viability of which was taken as 100%. The experiments with TMDF were performed 2 times with each sample tested in triplicates.

4.5. Morphological Studies

To determine morphological changes, cell cultures treated with various concentrations of TMDF or IM60 at different time-points and observed under an inverted light microscope attached to a charged couple device (CCD) camera or the EVOS Cell Imaging System (Thermo Fisher Scientific, Waltham, MA, USA). A comparison was performed between the treated cells and untreated cells or cells treated with similar DMSO concentrations.

4.6. Western Blot Analysis

Western blotting of protein lysates was performed to determine the mechanism of cell death of TMDF. The harvested cells treated with various concentrations of TMDF, DMSO, or untreated cells were washed with 1x phosphate-buffered saline (PBS) and lysed using 100 μL of radioimmunoprecipitation assay buffer (RIPA) lysis buffer (10 mM Tris-Cl [pH 8.0], 1 mM ethylenediaminetetraacetic acid (EDTA), 1% Triton X-100, 0.1% sodium deoxycholate, 0.1% sodium dodecyl sulfate (SDS), and 140 mM NaCl) per million cells supplemented with 50 μL of β-mercaptoethanol/mL RIPA and 1 mM of the serine protease inhibitor, phenylmethyl sulfonyl fluoride (PMSF). The lysed cells were spun at 14,000 rpm for 10 min at 4 °C to separate the nuclei, and the supernatants containing the cell extract either used immediately or stored at −80 °C. The quantification of the proteins in the cellular extracts was performed using the Bradford colorimetric assay (Bio-Rad Life Sciences, Hercules, CA, USA) as per manufacturer's instruction. Total cellular protein lysates (40 μg per lane) were loaded onto 8–12% SDS-polyacrylamide mini gels and blotted onto Protran nitrocellulose membranes (Whatman plc-GE Healthcare, Kent, UK) using standard protocol. The membranes were blocked with 5% low fat dried milk in 1× PBS and 0.1% Tween-20 (PBST) for 1 h and incubated at 4 °C overnight with 1:1000 dilution of various primary antibodies from Cell Signaling Technology (Danvers, MA, USA) against poly (ADP-ribose) polymerase (PARP), caspase 7, caspase 8, and caspase 9, or actin (Sigma-Aldrich, St. Louis, MO, USA) in 1% milk-PBST. The blots were then incubated with the appropriate secondary antibodies (anti-mouse or anti-rabbit) for 1 h and 30 min at room temperature, followed by detection using the Pierce™ ECL Plus Western Blotting Substrate (Thermo Fisher Scientific, Waltham, MA, USA). The chemiluminescent signal was captured using Typhoon FLA 9500 (GE Healthcare, Chicago, IL, USA).

4.7. Annexin V/Propidium Iodide Staining

Early events of apoptosis induction were studied using the classical Annexin V/PI staining assay employing flow cytometry. MCF-7 cells were treated with media alone, DMSO, or either 125 or 250 μg/mL of TMDF for six hours in duplicates. To inhibit apoptosis, cells were pretreated with 20 μM of Z-VAD-FMK (Promega Corporation, Fitchburg, WI, USA) before treatment with TMDF. Following treatment, cells were harvested with Accutase (Sigma-Aldrich, St. Louis, MO, USA), a trypsin substitute, and processed for flow cytometry using the Annexin V/PI kit from Becton Dickenson (Franklin Lakes, NJ, USA) as per manufacturer's instructions.

4.8. Statistical Analysis

Statistical analysis was performed using Microsoft Excel (Microsoft, Redmond, WA, USA) or GraphPad Prism version 5 (GraphPad Software, San Diego, CA, USA) or higher software. For each sample, averages were calculated in replicates of three or more, along with their standard deviations (plotted as error bars on the column graphs). Significance of variation between any two groups was assessed using paired, two-tailed student's *t*-test (* $p < 0.05$; ** $p < 0.01$ but >0.001; *** $p < 0.001$). Since a specific DMSO control was used for each tested extract/fraction, the dose- and time-dependent GraphPad figures were created by normalizing the values obtained for the test groups against their specific DMSO control.

5. Conclusions

This study characterized the anticancer potential of *T. mascatense* which resulted in the identification of a potential lead anticancer compound. We found that various organic extract/fractions of *T. mascatense*, especially TMDF, could induce cell death in breast cancer and cervical cell lines. In the breast cancer cells, the cell death was primarily caused by the induction of apoptosis, providing evidence to test its effectiveness and anti-proliferative activity in animal models. Furthermore, a lead sesquiterpene, IM60, was identified from TMDF with the potential to induce cell death in breast cancer cells. It would be valuable to characterize this compound experimentally further (or its derivatives) for its mode of action and conduct in silico docking analysis for breast cancer receptors to determine if it can be specifically targeted against breast cancer cells that are responsive to hormone treatment.

Author Contributions: Conceptualization, F.M.; Formal analysis, N.G.P., S.O.M.S.B., and F.M.; Funding acquisition, A.A.-H., J.H. and F.M.; Investigation, N.G.P., S.O.M.S.B., S.A., M.M.Q., T.S.R. and F.M.; Methodology, F.M.; Project administration, N.G.P. and F.M.; Resources, A.A.-H., J.H. and F.M.; Supervision, F.M.; Validation, N.G.P., S.O.M.S.B., and F.M.; Visualization, S.O.M.S.B., N.G.P., and F.M.; Writing—original draft, S.O.M.S.B. and F.M.; Writing—review & editing, J.H. and F.M.

Funding: This research was funded by the UAE University grant numbers 31M122 and 31R087 to F.M. and the Oman Research Council (TRC) Open Research Grant number ORG/CBS/12/004.

Acknowledgments: MCF-7 and MDA-MB-231 were obtained from Samir Attoub, Department of Pharmacology, College of Medicine and Health Sciences (CMHS), UAE University (UAEU). MCF-10A was obtained from ATCC (USA). HeLa cell line was a gift of Tahir A. Rizvi, Department of Microbiology and Immunology, CMHS, UAEU. We would like to thank Waqar Ahmad, Department of Biochemistry, CMHS, UAEU for his assistance in performing the IC_{50} calculations and with referencing.

Conflicts of Interest: The authors declare no conflict of interest.

References

1. WHO. Cancer: Key Facts. Geneva, Switzerland. Available online: http://www.who.int/mediacentre/factsheets/fs297/en/ (accessed on 1 February 2018).
2. GLOBOCAN 2018: Estimated Cancer Incidence, Mortality and Prevalence Worldwide in 2018. 2018. Available online: http://gco.iarc.fr/ (accessed in 1 September 2018).
3. Stewart, B.W.; Wild, C.P. (Eds.) *World Cancer Report 2014*; International Agency for Research on Cancer: Lyon, France, 2014.
4. Cragg, G.M.; Newman, D.J.; Yang, S.S. Natural product extracts of plant and marine origin having antileukemia potential. The NCI experience. *J. Nat. Prod.* **2006**, *69*, 488–498. [CrossRef] [PubMed]
5. Khan, H. Medicinal Plants in Light of History: Recognized Therapeutic Modality. *J. Evid. Based Complement. Altern. Med.* **2014**, *19*, 216–219. [CrossRef] [PubMed]
6. Verpoorte, R. Pharmacognosy in the New Millennium: Lead finding and Biotechnology. *J. Pharm. Pharmacol.* **2000**, *52*, 253–262. [CrossRef] [PubMed]
7. Levitsky, D.O.; Dembitsky, V.M. Anti-breast Cancer Agents Derived from Plants. *Nat. Prod. Bioprospect.* **2014**, *5*, 1–16. [CrossRef] [PubMed]
8. Tariq, A.; Sadia, S.; Pan, K.; Ullah, I.; Mussarat, S.; Sun, F.; Abiodun, O.O.; Batbaatar, A.; Li, Z.; Song, D.; et al. A systematic review on ethnomedicines of anti-cancer plants. *Phytother. Res.* **2017**, *31*, 202–264. [CrossRef] [PubMed]
9. Alves-Silva, J.M.; Romane, A.; Efferth, T.; Salgueiro, L. North African Medicinal Plants Traditionally Used in Cancer Therapy. *Front. Pharmacol.* **2017**, *8*, 383. [CrossRef] [PubMed]
10. Richards, T.S.; Kamdje, A.H.N.; Mukhtar, F. Medicinal Plants in Breast Cancer Therapy. *J. Dis. Med. Plants* **2015**, *1*, 19–23.
11. Western, A.R. *The Flora of United Arab Emirates, an Introduction*; Publication of the UAE University: Sharjah, UAE, 1986.
12. El-Ghonemy, A.A. *Encyclopedia of Medicinal Plants of the United Emirates*, 1st ed.; University of UAE: Sharjah, UAE, 1993.
13. Ghazanfar, S.A. *Handbook of Arabian Medicinal Plants*; Library of Congress: Washington, DC, USA, 1994.
14. Jongbloed, M.V. *The Comprehensive Guide to the Wild Flowers of the United Arab Emirates, ERWDA*; Emirates Printing Press: Dubai, UAE, 2003.
15. Miller, A.G.; Morris, M. *Plants of Dhofar, The southern Region of Oman: Traditional, Economic and Medicinal Uses*; Office of the Advisor for Conservation of the Environment: Muscat, Oman, 1987.
16. Zaragari, A. *Medicinal Plants*, 6th ed.; Tehran University Publications: Tehran, Iran, 1994.
17. Tutin, T.G.; Heywood, V.; Burges, N.A.; Moore, D.M.; Valentine, D.H.; Walters, S.M.; Webb, D.A. *Flora Europaea III. Genus teucrium L.*; Cambridge Press: Cambridge, UK, 1972.
18. Gharaibeh, M.N.; Elayan, H.H.; Salhab, A.S. Hypoglycemic effects of *Teucrium polium*. *J. Ethnopharmacol.* **1988**, *24*, 93–99. [CrossRef]
19. Rudolf-Pfaendler, H.; Piozzi, F.; Rodriguez, B.; Savona, G. Advances in the chemistry of the Furanoditerphenoids from Teucrium species. *Heterocycles* **1987**, *25*, 807–841. [CrossRef]
20. Suleiman, M.S.; Abdul-Ghani, A.S.; Al-Khalil, S.; Amin, R. Effect of *Teucrium polium* boiled leaf extract on intestinal motility and blood pressure. *J. Ethnopharmacol.* **1988**, *22*, 111–116. [CrossRef]
21. Stankovic, M.; Topuzovic, M.; Solujic, S.; Mihailovic, V. Antioxidant activity and concentration of phenols and flavonoids in the whole plant and plant parts of *Teucrium chamaerdys* L. var. glanduliferum Haussk. *J. Med. Plant Res.* **2010**, *4*, 2092–2098.
22. Shahraki, M.R.; Arab, M.R.; Mirimokaddam, E.; Palan, M.J. The effect of *Teucrium polium* (Calpoureh) on liver function, serum lipids and glucose in diabetic male rats. *Iran Biomed. J.* **2007**, *11*, 65–68. [PubMed]

23. Rasekh, H.R.; Khoshnood-Mansourkhani, M.J.; Kamalinejad, M. Hypolipidemic, Effects of *Teucrium polium* in rats. *Fitoterapia* **2001**, *72*, 937–939. [CrossRef]
24. Esmaeili, M.A.; Yazdanparast, R. Hypoglycaemic effect of *Teucrium polium*: Studies with rat pancreatic islets. *J. Ethnopharmacol.* **2004**, *95*, 27–30. [CrossRef] [PubMed]
25. Panovska, T.K.; Kulevanova, S.; Gjorgoski, I.; Bogdanova, M.; Petrushevska, G. Hepatoprotective effect of the ethyl acetate extract of *Teucrium polium* L. against carbon tetrachloride-induced hepatic injury in rats. *Acta Pharm.* **2007**, *57*, 241–248. [CrossRef] [PubMed]
26. Tariq, M.; Ageel, A.M.; al-Yahya, M.A.; Mossa, J.S.; al-Said, M.S. Anti-inflammatory activity of *Teucrium polium*. *Int. J. Tissue React.* **1989**, *11*, 185–188. [PubMed]
27. Heidari, M.; Karaminezhad, R.; Davvand, E.; Jalali, S. Evaluation of the analgesic effect of *Teucrium polium* extract in mice. *DARU J. Pharm. Sci.* **2001**, *24*, 277–285.
28. Kadifkova Panovska, T.; Kulevanova, S.; Stefova, M. In vitro antioxidant activity of some Teucrium species (Lamiaceae). *Acta Pharm.* **2005**, *55*, 207–214. [PubMed]
29. Hasani, P.; Yasa, N.; Vosough-Ghanbari, S.; Mohammadirad, A.; Dehghan, G.; Abdollahi, M. In vivo antioxidant potential of *Teucrium polium*, as compared to alpha-tocopherol. *Acta Pharm.* **2007**, *57*, 123–129. [CrossRef] [PubMed]
30. Jurisic, R.; Vladimir-Knezevic, S.; Kalodera, Z.; Grgic, J. Determination of selenium in Teucrium species by hydride generation atomic absorption spectrometry. *Z. Naturforsch. C* **2003**, *58*, 143–145. [CrossRef] [PubMed]
31. Stanković, M.; Topuzović, M.; Marković, A.; Pavlovic, D.; Solujic, S.; Niciforovic, N.; Mihailovic, V. Antioxidant Activity, Phenol and flavonoid contents of different *Teucrium chamaedrys* L. extracts. *Biotechnol Equip* **2010**, *24*, 82–86. [CrossRef]
32. Camarda, I. Ricerche etnobotaniche nel comune di Dorgali (Sardegna centro-orientale). *Bollettino della Società Sarda di Scienze Naturali* **1990**, *27*, 147–204.
33. Bahramikia, S.; Yazdanparast, R. Phytochemistry and medicinal properties of *Teucrium polium* L. (Lamiaceae). *Phytother. Res.* **2012**, *26*, 1581–1593. [CrossRef] [PubMed]
34. Rajabalian, S. Methanolic extract of *Teucrium polium* L. potentiates the cytotoxic and apoptotic effects of anticancer drugs of vincristine, vinblastine and doxorubicin against a panel of cancerous cell lines. *Exp. Oncol.* **2008**, *30*, 133–138. [PubMed]
35. Bedir, E.; Tasdemir, D.; Çalis, I.; Zerbe, O.; Sticher, O. Neo-clerodane diterpenoids from *Teucrium polium*. *Phytochemistry* **1999**, *51*, 921–925. [CrossRef]
36. Piozzi, F. Further researches on the furoclerodanes from Teucrium species. *Heterocycles* **1994**, *37*, 603–626. [CrossRef]
37. Piozzi, F.; Bruno, M.; Rosselli, S. Further furoclerodanes from Teucrium genus. *Heterocycles* **1998**, *48*, 2185–2203. [CrossRef]
38. Acamovic, T.; Brooker, J.D. Biochemistry of plant secondary metabolites and their effects in animals. *Proc. Nutr. Soc.* **2005**, *64*, 403–412. [CrossRef] [PubMed]
39. Kandouz, M.; Alachkar, A.; Zhang, L.; Dekhil, H.; Chehna, F.; Yasmeen, A.; Al Moustafa, A.E. *Teucrium polium* plant extract inhibits cell invasion and motility of human prostate cancer cells via the restoration of the E-cadherin/catenin complex. *J. Ethnopharmacol.* **2010**, *129*, 410–415. [CrossRef] [PubMed]
40. Movahedi, A.; Basir, R.; Rahmat, A.; Charaffedine, M.; Othman, F. Remarkable Anticancer Activity of *Teucrium polium* on Hepatocellular Carcinogenic Rats. *Evid. Based Complement. Altern. Med.* **2014**, *2014*, 726724. [CrossRef] [PubMed]
41. Rehman, N.; Al-Sahai, J.M.S.; Hussain, J.; Khan, A.L.; Gilani, S.A.; Abbas, G.; Hussain, J.; Sabahi, J.N.; Al-Harassi, A. Phytochemical and pharmacological investigation of *Teucrium muscatense*. *Int. J. Phytomed.* **2016**, *8*, 567–579.
42. Eloff, J.N. Which extractant should be used for the screening and isolation of antimicrobial components from plants? *J. Ethnopharmacol.* **1998**, *60*, 1–8. [CrossRef]
43. Soule, H.D.; Vazguez, J.; Long, A.; Albert, S.; Brennan, M. A human cell line from a pleural effusion derived from a breast carcinoma. *J. Natl. Cancer Inst.* **1973**, *51*, 1409–1416. [CrossRef] [PubMed]
44. Scherer, W.F.; Syverton, J.T.; Gey, G.O. Studies on the propagation in vitro of poliomyelitis viruses. IV. Viral multiplication in a stable strain of human malignant epithelial cells (strain HeLa) derived from an epidermoid carcinoma of the cervix. *J. Exp. Med.* **1953**, *97*, 695–710. [CrossRef] [PubMed]

45. Soule, H.D.; Maloney, T.M.; Wolman, S.R.; Peterson, W.D., Jr.; Brenz, R.; McGrath, C.M.; Russo, J.; Pauley, R.J.; Jones, R.F.; Brooks, S.C. Isolation and characterization of a spontaneously immortalized human breast epithelial cell line, MCF-10. *Cancer Res.* **1990**, *50*, 6075–6086. [PubMed]
46. Cailleau, R.; Young, R.; Olive, M.; Reeves, W.J., Jr. Breast tumor cell lines from pleural effusions. *J. Natl. Cancer Inst.* **1974**, *53*, 661–674. [CrossRef] [PubMed]
47. Wahba, H.A.; El-Hadaad, H.A. Current approaches in treatment of triple-negative breast cancer. *Cancer Biol. Med.* **2015**, *12*, 106–116. [PubMed]
48. Iqbal, J.; Abbasi, B.A.; Mahmood, T.; Kanwal, S.; Ali, B.; Shah, S.A.; Khalil, A.T. Plant-derived anticancer agents: A green anticancer approach. *Asian Pac. J. Trop. Biomed.* **2017**, *7*, 1129–1150. [CrossRef]
49. Elmore, S. Apoptosis: A review of programmed cell death. *Toxicol. Pathol.* **2007**, *35*, 495–516. [CrossRef] [PubMed]
50. Desouza, M.; Gunning, P.W.; Stehn, J.R. The actin cytoskeleton as a sensor and mediator of apoptosis. *Bioarchitecture* **2012**, *2*, 75–87. [CrossRef] [PubMed]
51. Stennicke, H.R.; Salvesen, G.S. Biochemical characteristics of caspases-3, -6, -7, and -8. *J. Biol. Chem.* **1997**, *272*, 25719–25723. [CrossRef] [PubMed]
52. Suhail, M.M.; Wu, W.; Cao, A.; Mondalek, F.G.; Fung, K.M.; Shih, P.T.; Fang, Y.T.; Woolley, C.; Young, G.; Lin, H.K. Boswellia sacra essential oil induces tumor cell-specific apoptosis and suppresses tumor aggressiveness in cultured human breast cancer cells. *BMC Complement. Altern. Med.* **2011**, *11*, 129. [CrossRef] [PubMed]
53. Janicke, R.U. MCF-7 breast carcinoma cells do not express caspase-3. *Breast Cancer Res. Treat.* **2009**, *117*, 219–221. [CrossRef] [PubMed]
54. Gali-Muhtasib, H.; Hmadi, R.; Kareh, M.; Tohme, R.; Darwiche, N. Cell death mechanisms of plant-derived anticancer drugs: Beyond apoptosis. *Apoptosis* **2015**, *20*, 1531–1562. [CrossRef] [PubMed]
55. Wink, M.; Alfermann, A.W.; Frank, R.; Wetterauer, B.; Disti, M.; Windhovel, J.; Krohn, O.; Fuss, E.; garden, H.; Mohagheghzadeh, A.; et al. Sustainable bioproduction of phytochemicals by plant in vitro cultures: Anticancer agents. *Plant Genet. Resour.* **2005**, *3*, 90–100. [CrossRef]
56. Newman, D.J.; Cragg, G.M.; Snader, K.M. Natural products as sources of new drugs over the period 1981–2002. *J. Nat. Prod.* **2003**, *66*, 1022–1037. [CrossRef] [PubMed]
57. Ali, L.; Hussain, I.; Rizvi, T.S.; Khan, A.L.; Shaukat, S.; Al-Sahi, J.M.S.; Al-Harrasi, A.; Hussain, J. A New Irregular Trihydroxy Sesquiterpene from *Teucrium mascatense*. *Helv. Chim. Acta* **2015**, *98*, 1462–1465. [CrossRef]
58. Stankovic, M.S.; Curcic, M.G.; Zizic, J.B.; Topuzovic, M.D.; Solujic, S.R.; Markovic, S.D. Teucrium plant species as natural sources of novel anticancer compounds: Antiproliferative, proapoptotic and antioxidant properties. *Int. J. Mol. Sci.* **2011**, *12*, 4190–4205. [CrossRef] [PubMed]
59. Tafrihi, M.; Toosi, S.; Minaei, T.; Gohari, A.R.; Niknam, V.; Arab Najafi, S.M. Anticancer properties of Teucrium persicum in PC-3 prostate cancer cells. *Asian Pac. J. Cancer Prev.* **2014**, *15*, 785–791. [CrossRef] [PubMed]
60. Jaradat, N.; Al-Lahham, S.; Abualhasan, M.N.; Bakri, A.; Zaide, H.; Hammad, J.; Hussein, F.; Issa, L.; Mousa, A.; Speih, R. Chemical Constituents, Antioxidant, Cyclooxygenase Inhibitor, and Cytotoxic Activities of Teucrium pruinosum Boiss. Essential Oil. *BioMed Res. Int.* **2018**, *2018*, 4034689. [CrossRef] [PubMed]
61. Cowan, M.M. Plant products as antimicrobial agents. *Clin. Microbiol. Rev.* **1999**, *12*, 564–582. [CrossRef] [PubMed]
62. Ncube, N.; Afolayan, A.; Okoh, A. Assessment techniques of antimicrobial properties of natural compounds of plant origin: Current methods and future trends. *Afr. J. Biotechnol.* **2008**, *7*, 1797–1806. [CrossRef]
63. Chaves, A.L.; Vergara, C.E.; Mayer, J.E. Dichloromethane as an economic alternative to chloroform in the extraction of DNA from plant tissues. *Plant Mol. Biol. Rep.* **1995**, *13*, 18–25. [CrossRef]
64. Al-Oqail, M.M.; Al-Sheddi, E.S.; Siddiqui, M.A.; Musarrat, J.; Al-Khedhairy, A.A.; Farshori, N.N. Anticancer Activity of Chloroform Extract and Sub-fractions of Nepeta deflersiana on Human Breast and Lung Cancer Cells: An In vitro Cytotoxicity Assessment. *Pharmacogn. Mag.* **2015**, *11* (Suppl. 4), S598–S605. [PubMed]
65. Kh, A.H.; Kh, S.; Yaseen, N. Evaluation the activity of chloroform extract of *Solanum nigrum* on non-Hodgkin lymphoma cell line (SR), rat embryo fibroblast cell line (REF) and human lymphocytes in vitro. *World J. Pharm. Sci.* **2015**, *4*, 14–22.

66. Modzelewska, A.; Sur, S.; Kumar, S.K.; Khan, S.R. Sesquiterpenes: Natural products that decrease cancer growth. *Curr. Med. Chem. Anticancer Agents* **2005**, *5*, 477–499. [CrossRef] [PubMed]
67. Bartikova, H.; Hanusova, V.; Skalova, L.; Ambroz, M.; Bousova, I. Antioxidant, pro-oxidant and other biological activities of sesquiterpenes. *Curr. Top. Med. Chem.* **2014**, *14*, 2478–2494. [CrossRef] [PubMed]
68. Li, G.; Kusari, S.; Spiteller, M. Natural products containing 'decalin' motif in microorganisms. *Nat. Prod. Rep.* **2014**, *31*, 1175–1201. [CrossRef] [PubMed]
69. Menichini, F.; Conforti, F.; Rigano, D.; Formisano, C.; Piozzi, F.; Senatore, F. Phytochemical composition, anti-inflammatory and antitumour activities of four Teucrium essential oils from Greece. *Food Chem.* **2009**, *115*, 679–686. [CrossRef]
70. Guaman-Ortiz, L.M.; Orellana, M.I.; Ratovitski, E.A. Natural Compounds as Modulators of Non-apoptotic Cell Death in Cancer Cells. *Curr. Genom.* **2017**, *18*, 132–155. [CrossRef] [PubMed]
71. Chen, N.; Karantza, V. Autophagy as a therapeutic target in cancer. *Cancer Biol. Ther.* **2011**, *11*, 157–168. [CrossRef] [PubMed]
72. Dai, X.; Cheng, H.; Bai, Z.; Li, J. Breast Cancer Cell Line Classification and Its Relevance with Breast Tumor Subtyping. *J. Cancer* **2017**, *8*, 3131–3141. [CrossRef] [PubMed]
73. Sulaiman, A.; Wang, L. Bridging the divide: Preclinical research discrepancies between triple-negative breast cancer cell lines and patient tumors. *Oncotarget* **2017**, *8*, 113269–113281. [CrossRef] [PubMed]
74. Suffredini, I.B.; Paciencia, M.L.; Frana, S.A.; Varella, A.D.; Younes, R.N. In vitro breast cancer cell lethality of Brazilian plant extracts. *Pharmazie* **2007**, *62*, 798–800. [PubMed]
75. Fouche, G.; Cragg, G.M.; Pillay, P.; Kolesnikova, N.; Maharaj, V.J.; Senabe, J. In vitro anticancer screening of South African plants. *J. Ethnopharmacol.* **2008**, *119*, 455–461. [CrossRef] [PubMed]
76. Riss, T.L.; Moravec, R.A.; Niles, A.L.; Duellman, S.; Benink, H.A.; Worzella, T.J.; Minor, L. Cell Viability Assays. In *Assay Guidance Manual*; Sittampalam, G.S., Coussens, N.P., Brimacombe, K., Grossman, A., Arkin, M., Auld, D., Austin, C., Baell, J., Bejcek, B., Caaveiro, J.M.M., et al., Eds.; Eli Lilly & Company and the National Center for Advancing Translational Sciences: Bethesda, MD, USA, 2004.

Sample Availability: Samples of TMDF may be available from the authors, if not exhausted.

molecules

MDPI

Article

Reduction of Preneoplastic Lesions Induced by 1,2-Dimethylhydrazine in Rat Colon by Maslinic Acid, a Pentacyclic Triterpene from *Olea europaea* L.

M. Emília Juan [1,2,*,†], Glòria Lozano-Mena [1,2,†], Marta Sánchez-González [1,2] and Joana M. Planas [1,2,*]

1 Departament de Bioquímica i Fisiologia, Universitat de Barcelona (UB), Av. Joan XXIII 27-31, 08028 Barcelona, Spain; gloria.lozano.mena@gmail.com (G.L.-M.); chan_marta@hotmail.com (M.S.-G.)

2 Institut de Recerca en Nutrició i Seguretat Alimentària (INSA), Universitat de Barcelona (UB), Av. Prat de la Riva 171, 08921 Santa Coloma de Gramanet, Spain

* Correspondence: mejuan@ub.edu (M.E.J.); jmplanas@ub.edu (J.M.P.); Tel.: +34-93-402-4505 (M.E.J. & J.M.P.)

† M.E.J. and G.L.-M. contributed equally to the article.

Academic Editors: Eva E. Rufino-Palomares and José Antonio Lupiáñez

Received: 6 March 2019; Accepted: 29 March 2019; Published: 1 April 2019

check for updates

Abstract: Maslinic acid triggers compelling antiproliferative and pro-apoptotic effects in different human cancer cell lines. Hence, the chemopreventive activity was investigated on early stages of carcinogenesis induced by 1,2-dimethylhydrazine (DMH) which is a model that mimics human sporadic colorectal cancer. Male Sprague-Dawley rats were orally administered either maslinic acid at 5, 10 or 25 mg/kg dissolved in (2-hydroxypropyl)-β-cyclodextrin 20% (*w*/*v*) or the solvent for 49 days. After one week of treatment, animals received three weekly intraperitoneal injections of DMH at the dose of 20 mg/kg. Maslinic acid reduced the preneoplastic biomarkers, aberrant crypt foci (ACF) and mucin-depleted foci (MDF), already at 5 mg/kg in a 15% and 27%, respectively. The decline was significant at 25 mg/kg with decreases of 33% and 51%, respectively. Correlation analysis showed a significant association between the concentrations of maslinic acid found in the colon and the reduction of ACF (r = 0.999, p = 0.019) and MDF (r = 0.997, p = 0.049). The present findings demonstrate that maslinic acid induced an inhibition of the initiation stages of carcinogenesis. The assessment of this pentacyclic triterpene at the colon sheds light for designing diets with foods rich in maslinic acid to exert a chemopreventive activity in colorectal cancer.

Keywords: aberrant crypt foci; colon cancer; mucin depleted foci; maslinic acid; dimethylhydrazine

1. Introduction

Colorectal cancer (CRC) is one of the most frequent malignancies in the world, ranking third in incidence and second in mortality in both genders [1]. Epidemiological studies support that lifestyle along with nutritional interventions can prevent CRC and that Mediterranean diet is associated with a lower prevalence of this disease in humans of all races [2–4]. This eating pattern is characterized by a high consumption of fruits, vegetables and legumes that provides a plethora of bioactive compounds with antitumoral activity [2–4]. Among them, stand out resveratrol contained in grapes [5], sulforaphane in cruciferous vegetables [6] or luteolin in celery and parsley [7]. Hence, chemoprevention with natural compounds is currently proposed as a dietary strategy for the control and constraint of carcinogenesis [8]. Therefore, here we focus our attention on maslinic acid, also known as crategolic acid or 2α,3β-dihydroxyolean-12-en-28-oic acid (Figure 1) with molecular weight of 472.7 g/mol. This phytochemical belongs to the group of pentacyclic triterpenes and is broadly distributed in nature, being found in different foods such as olives, spinaches, eggplants, chickpeas, lentils, kiwi and pomegranates as well as in plants used in traditional Asian medicine for the treatment of different affections [9]. In this sense, the leaves of banaba or *Lagerstroemia speciosa* L. have been used for the treatment of diabetes, *Crataegus monogyna* L. commonly known as hawthorn, constitute a remedy for cardiovascular diseases, and the leaves of loquat or *Eriobotrya japonica* L. have been employed as antitussive and anti-inflammatory for chronic bronchitis [9].

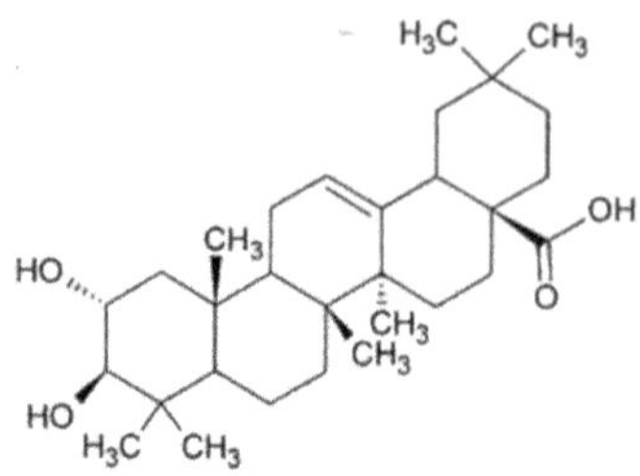

Figure 1. Chemical structure of maslinic acid.

Maslinic acid is gaining interest due to its lack of harmful effects [10] together with its multiple beneficial activity on health, essentially antidiabetic, antioxidant, anti-inflammatory, cardioprotective, neuroprotective and antitumoral [9]. Among them, it its chemopreventive activity in the human colon adenocarcinoma cell lines HT29 and Caco-2 cells [11–13] is worth mentioning. Moreover, it was also demonstrated that after the oral administration of maslinic acid, high concentrations were found in the colon due to its low bioavailability [14]. Taking together these results, they suggest that the distal part of the intestine could be a target organ where maslinic acid can exert its beneficial effects.

Although the promising activities demonstrated in vitro, the antitumoral action in colon cancer in vivo has only been evaluated in male $Apc^{Min/+}$ mice which is a genetic model that mimics the familial cancers such as human familial adenomatous polyposis (FAP) and hereditary non-polyposis colon cancer. Results indicated that maslinic acid at 100 mg/kg of diet suppressed polyp formation by a 54% [15]. However, the most common cancer is the non-familial colorectal type, which occurs sporadically and can be induced in animal models with the use of carcinogens [16]. Therefore, the chemopreventive effect of maslinic acid was investigated on colonic preneoplastic lesions induced by the administration of 1,2-dimethylhydrazine (DMH) in rats. To that end, maslinic acid was administered at 5, 10 and 25 mg/kg, being the lower dose easily attained following a Mediterranean dietary pattern. Maslinic acid efficacy was assessed in terms of appearance of aberrant crypt foci (ACF) that were determined as biomarker of cancer risk and mucin depleted foci (MDF) measured as biomarker of dysplasia since they harbor characteristics comparable to microadenomas [17]. The reduction of these preneoplastic biomarkers by the phytochemical will confirm its action in vivo. In addition, the relationship between the colonic concentrations of maslinic acid and the occurrence of ACF and MDF were assessed. For that purpose, maslinic acid was quantified in the colon content by LC-APCI-MS analysis and established the correlation between its concentrations in the colon and the reduction of preneoplastic biomarkers. The results provide new insights not only on the maslinic effects on gut health but also a thorough basis to allow a dietary recommendation of the intake of foods with a high content of this pentacyclic triterpene.

2. Results

2.1. Body Weight, Food and Water Consumption, and Food Conversion Efficiency

Careful observation of the animals during the experimental period showed no mortality or adverse effects. The consistency of stools was pelleted and firm, with no visible differences throughout the seven groups.

Body weight was not affected by the three consecutive intraperitoneal injections of DMH compared to the controls. Moreover, the daily oral administration of maslinic acid at the doses of 5, 10 or 25 mg/kg for 49 days did not modify body weight with respect to the control groups (Figure 2).

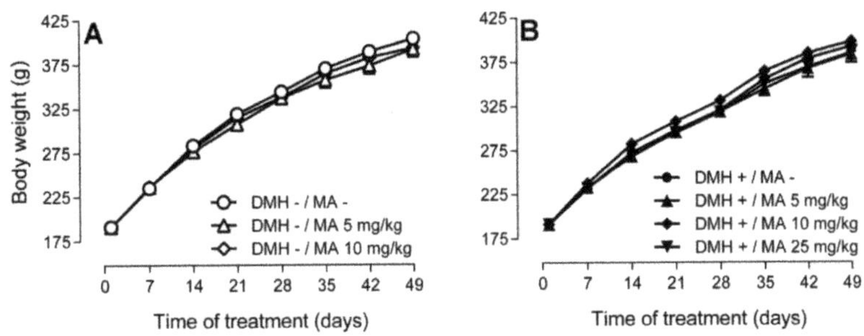

Figure 2. Body weight of Sprague-Dawley rats (**A**) that did not receive the carcinogen and were orally administered with solvent or maslinic acid at 5 or 10 mg/kg, whereas (**B**) displays the groups that received 1,2-dimethylhydrazine (DMH) once a week for three weeks long with solvent or maslinic acid at 5, 10 or 25 mg/kg. Results are expressed as mean ± standard error of the mean (SEM) (n = 6–8) and were analyzed by two-way ANOVA, followed by Tukey's multiple comparisons test. No significant differences ($p > 0.05$) were found between groups submitted at the different treatments.

The seven groups did not differ in food and water consumption (data not shown). Figure 3 displays the food conversion efficiency (FCE) that showed the same pattern in all experimental groups. FCE was highest during the first week, decreased during the second and third week, and remained constant thereafter, with no significant differences between groups.

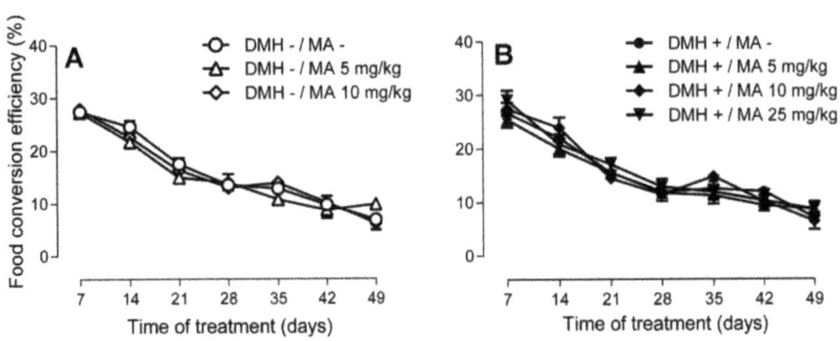

Figure 3. Food conversion efficiency of Sprague-Dawley rats (**A**) that were not intraperitoneally injected with the carcinogen and were orally administered with solvent or maslinic acid at 5 or 10 mg/kg, whereas (**B**) depicts the groups that received DMH once a week for three weeks along with solvent or maslinic acid at 5, 10 or 25 mg/kg. Results are expressed as mean ± SEM (n = 6–8) and were analyzed by two-way ANOVA, followed by Tukey's multiple comparisons test. No significant differences ($p > 0.05$) were found between groups at the different treatments. Differences over time: DMH−/MA−, DMH−/MA 10, DMH+/MA10, 7 d = 14 d > 21 d = 28 d = 35 d = 42 d = 49 d; DMH+/MA−; DMH−/MA 5, DMH+/MA5, DMH+/MA25, 7 d > 14 d > 21 d = 28 d = 35 d = 42 d = 49 d.

2.2. *Aberrant Crypt Foci*

The colon mucosa of rats in the three DMH—groups, namely, intraperitoneally injected with EDTA 1 mmol/L (pH 6.5) and orally administered with either (2-hydroxypropyl)-β-cyclodextrin 20%

(*w*/*v*) or 5 and 10 mg/kg of maslinic acid did not show any microscopically observable alterations compatible with the presence of ACF. Given that no preneoplastic lesions were found in the three groups that were injected with saline solution (Figure 4A), all subsequent analysis were restricted to the four groups that received DMH and developed ACF (Figure 4B–D).

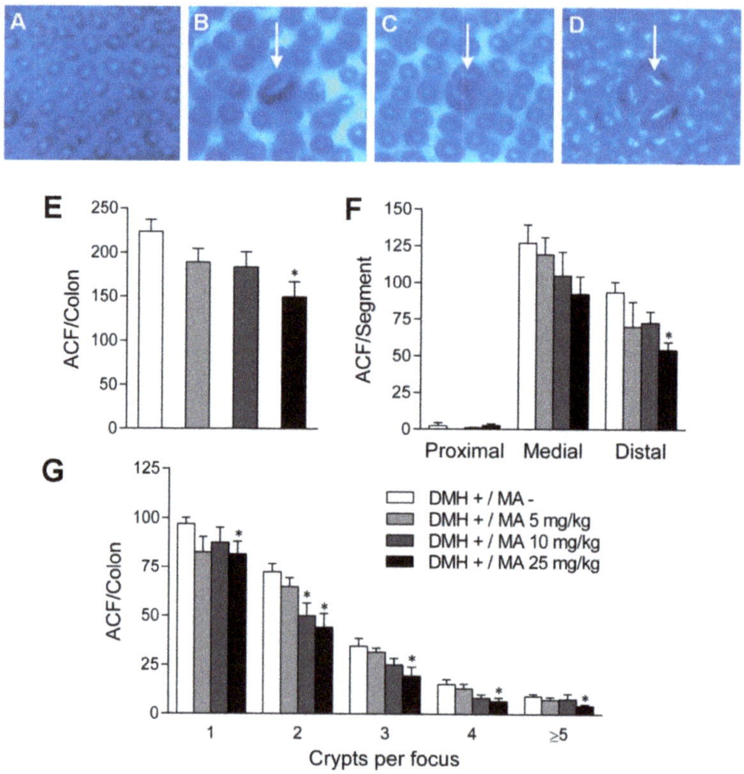

Figure 4. Aberrant crypt foci (ACF) observed under a light microscope after staining of the colon with methylene blue (magnification of ×10). The images show the whole mount colon of animals in (**A**) the negative control (DMH−/MA−) and the positive control (DMH−/MA+) showing a topographic view of ACF indicated by a white arrow, with one (**B**), two (**C**) or three crypts (**D**). The effects of maslinic acid at the doses of 0, 5, 10 and 25 mg/kg on the number of ACF in (**E**) total colon; (**F**) colonic segments and (**G**) number of crypt per focus in total colon. Results are expressed as mean ± SEM and were analyzed by the non-parametric Kruskal–Wallis test, followed by Dunn's multiple comparisons test, n = 6–8. Asterisks indicate different from the control group: * $p < 0.05$.

The number of preneoplastic lesions in colon was higher in the DMH+/MA− group (224 ± 37 ACF), than in the maslinic acid treated animals. This pentacyclic triterpene reduced the lesions to 189 ± 40 ($p > 0.05$), 184 ± 38 ($p > 0.05$) and 150 ± 43 ($p < 0.05$), at the doses of 5, 10 and 25 mg/kg, respectively (Figure 4E). ACF followed a regional distribution along the colon that was similar in all groups. In the positive control, ACF were practically absent in the proximal segment, increased to 127.1 ± 12.4 in the medial segment and were 98.4 ± 6.32 in the distal part (Figure 4F). Maslinic acid diminished the formation of ACF at the three doses evaluated, being 25 mg/kg the most effective one with a count of 92.3 ± 12.1 in the medial ($p > 0.05$) and 54.5 ± 5.1 in the distal segments ($p < 0.05$). Noteworthy, at the three doses evaluated maslinic acid was more active in inhibiting ACF in the distal parts of the colon, with reductions of 25% ($p > 0.05$), 22% ($p > 0.05$) and 42% ($p < 0.05$) at 5, 10 and 25 mg/kg, respectively.

The number of AC in each focus or crypt multiplicity was also determined (Figure 4G). In the DMH+/MA− group, most of the lesions were formed by a single crypt (42%), followed by the number of foci containing 2 crypts (32%), while thereafter, the foci comprising 3, 4 and ≥5 crypts were progressively lower, appearing at a 15%, 7% and 4%, respectively. A similar distribution was observed in the treated groups, in which maslinic acid reduced the number of lesions (Figure 4G). The effect exerted in the foci with higher number of crypts, since those with 2 crypts decreased in a 31% ($p < 0.05$) and 39% ($p < 0.05$) at the doses of 10 and 25 mg/kg, respectively. This pattern was also repeated at the foci with 3 crypts that dropped in a 27% ($p > 0.05$) and 43% ($p < 0.05$) at 10 and 25 mg/kg, respectively. Remarkable is that maslinic acid also induced a significant decrease in the foci with 4 and ≥5 crypts at the dose of 25 mg/kg since these lesions were reduced in a 56% ($p < 0.05$) and 54%, respectively ($p < 0.05$).

Aberrant crypts (AC) in the colon of the animals that were injected intraperitoneally with DMH but did not receive maslinic acid were 445 ± 36. The oral administration of this compound reduced AC to 388 ± 31 ($p > 0.05$), 341 ± 42 ($p > 0.05$) and 258 ± 42 ($p < 0.01$), at the doses of 5, 10 and 25 mg/kg, respectively.

2.3. Mucin-Depleted Foci

The oral administration of maslinic acid prevented the formation of dysplastic lesions or MDF which are aberrant crypts characterized by the loss of mucin production (Figure 5C,E). While in the DMH+/MA− group the number of MDF was 45.3 ± 4.0, the incidence decreased to 32.9 ± 6.6 ($p > 0.05$), 30.4 ± 8.5 ($p > 0.05$) and 22.3 ± 7.8 ($p < 0.05$), in the groups treated with 5, 10 and 25 mg/kg, respectively (Figure 5F). The depletion of mucins in aberrant crypts followed the same regional distribution observed for ACF (Figure 5G). Maslinic acid groups had fewer MDF than the DMH group, with the highest efficiency at the dose of 25 mg/kg.

The incidence of crypt multiplicity for 1, 2 and 3 crypts per focus was markedly lower in the maslinic acid groups than in the DMH+/MA− group (Figure 5H). Total mucin depleted crypts in the DMH+/MA− group were 105.4 ± 7.3, that were reduced to 78.1 ± 16.7 ($p > 0.05$), 61.2 ± 17.0 ($p > 0.05$) and 49.5 ± 18.1 ($p < 0.05$), in the groups that were orally administered with the doses of 5, 10 and 25 mg/kg of maslinic acid. Therefore, this bioactive compound was able to decrease the number of dysplastic mucin depleted aberrant crypts (MDAC) in colon by 26%, 42% and 53%, at the doses of 5, 10 and 25 mg/kg of maslinic acid.

2.4. Determination of AST and ALT

The hepatic enzymes were determined in order to assess the safety of experimental animals. Therefore, AST and ALT were analyzed in the three representative groups: the control rats, that did not receive neither the treatment, nor maslinic acid (DMH−/MA−), the animals challenged with the carcinogen that only received the solvent (DMH+/MA−), and the ones administered with the highest dose of maslinic acid at 25 mg/kg (DMH+/MA 25). The results obtained for AST and ALT indicate that the hepatic integrity of the animals is maintained throughout the experiment, since no significant differences were observed between groups (Table 1).

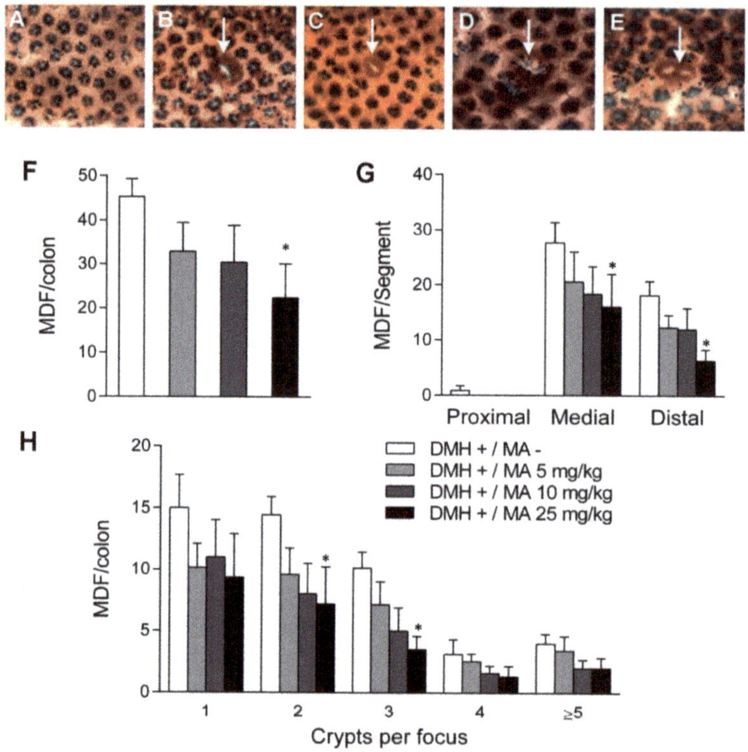

Figure 5. Mucin depleted foci (MDF) assessed after dying the colon with high-iron diamine/alcian blue/neutral red staining (HID-AB) and analyzed under a light microscope (magnification of ×10). Topographic features of the colonic epithelium of animals in (**A**) the negative control (DMH−/MA−) and the positive control (DMH+/MA−) displaying ACF with one (**B**) or two (**D**) crypts and MDF with one (**C**) or two crypts (**E**). The effects of maslinic acid at the doses of 0, 5, 10 and 25 mg/kg on the number of MDF in (**F**) total colon; (**G**) colonic segments and (**H**) number of crypt devoid of mucin per focus in total colon. Results are expressed as mean ± SEM and were analyzed by the non-parametric Kruskal–Wallis test, followed by Dunn's multiple comparisons test, n = 6–8. Asterisks indicate different from the control group: * $p < 0.05$.

Table 1. Hepatic enzymes in animals in the negative control group, positive control group and injected with DMH and MA at 25 mg/kg treatment group [1].

	DMH−/MA−	DMH+/MA−	DMH+/MA 25
AST, *UI/L*	161 ± 20 (n = 8)	172 ± 30 (n = 8)	178 ± 15 (n = 6)
ALT, *UI/L*	54.8 ± 2.7 (n = 8)	53.5 ± 3.9 (n = 8)	54.1 ± 2.8 (n = 6)

[1] Values are means ± SEM and were evaluated with one-way ANOVA and Bonferroni's multiple comparisons test. No significant differences ($p > 0.05$) were observed between groups.

2.5. Quantification of Maslinic Acid in Colon Content

Figure 6 depicts the typical chromatograms of the colon content from rats challenged with DMH that received the solvent or maslinic acid at the dose of 10 mg/kg for 49 days obtained between 18 and 20 h after the last oral administration.

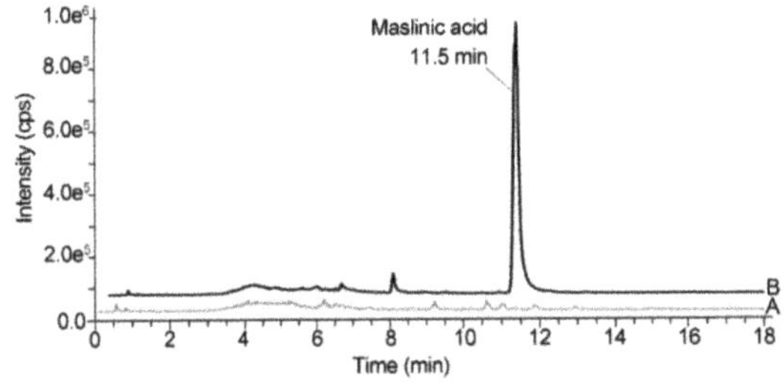

Figure 6. Representative liquid chromatography-mass spectrometry (LC-MS) chromatograms obtained in single ion monitoring (SIM) acquisition mode at m/z 471.3 of the colon content of (A) a rat from the DMH+/MA− group and (B) an animal from the DMH+/MA 10 mg/kg group.

The colon content of the rats in the DMH+/MA− group did not show any maslinic acid, whereas the animals orally administered with the pentacyclic triterpene showed a peak at 11.5 min that eluted free from other interfering peaks, thus indicating the selectivity of the analytical method.

Maslinic acid was found in the colon content achieving concentrations of 18.4 ± 2.7 nmol/g; 90.7 ± 36.9 nmol/g and 453.6 ± 114.3 nmol/g in animals that received 5, 10 and 25 mg/kg (n = 6), respectively. The DMH+/MA− group was also analyzed, and no maslinic acid was found in any sample (Figure 7).

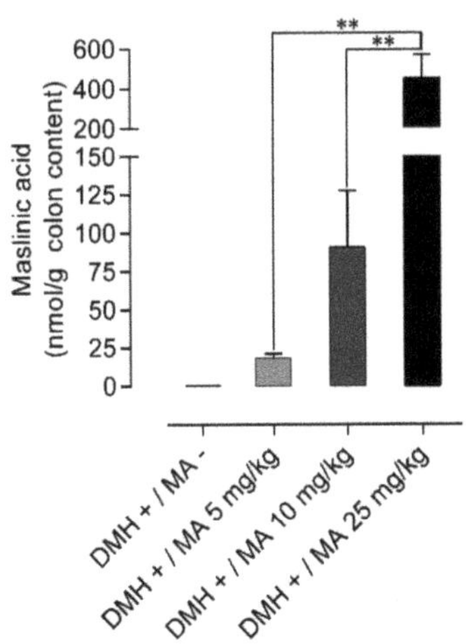

Figure 7. Concentrations of maslinic acid in the colon content of DMH-injected rats obtained 24 h after the last oral administration of the pentacyclic triterpene at the doses of 0, 5, 10 and 25 mg/kg after 49 days of treatment. Results are expressed as mean + SEM and were evaluated with one-way ANOVA and Bonferroni's multiple comparisons test, n = 6–8. Asterisks indicate differences between doses: ** $p < 0.01$.

The possible effect of DMH treatment on the concentrations of the pentacyclic triterpene that reached the colon was evaluated. Therefore, the content of the rats in the group that was not injected with the carcinogen but was orally administered throughout the experimental period with the pentacyclic triterpene at the doses of 10 mg/kg was analyzed. The concentration of maslinic acid in the colon was 86.4 ± 7.6 nmol/g (n = 6), which did not differ from the 90.7 ± 36.9 nmol/g achieved in the DMH+/MA 10 mg/kg group ($p > 0.05$). Given that DMH treatment did not modify the amounts of maslinic acid reaching the colon, the animals in the DMH−/MA 5 mg/kg were not included in the analysis.

2.6. Correlation between the Reductions of Preneoplastic Markers and Concentrations in the Colonic Content

The concentrations of maslinic acid found in the colon content were determined to establish a correlation between the biological effects observed in the colon and the amounts of the bioactive compound in the site of action (Figure 8). The percentage of reduction of both, ACF and MDF versus the amount of maslinic acid in the colon content showed a strong uphill linear relationship with Pearson's correlation coefficients higher than 0.99. The activity of maslinic acid in reducing preneoplastic lesions was parallel in the three doses studied but the effect of this pentacyclic triterpene was approximately 1.8-folds higher on reducing MDF than ACF.

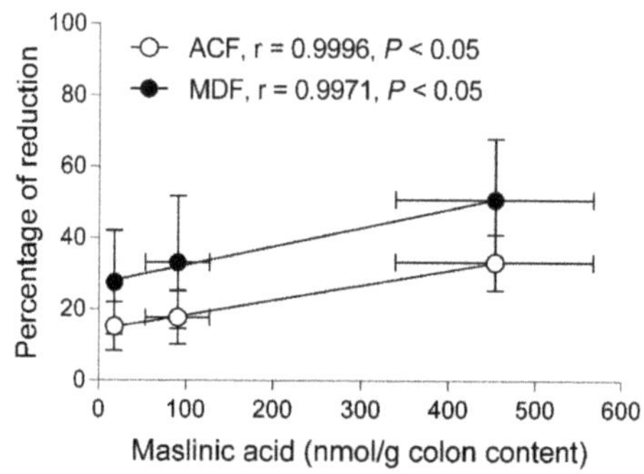

Figure 8. Pearson correlations of the reductions of aberrant crypt foci (ACF) and mucin depleted foci (MDF) versus concentrations of maslinic acid in the colonic content. Values are represented as means ± SEM. Horizontal bars are the SEM of the concentrations of maslinic acid in the colon (*n* = 6 for each dose). Vertical bars represent the SEM of the percentage of reduction of ACF (*n* = 6 for each dose) or MDF (*n* = 6 for each dose).

3. Discussion

The effect of maslinic acid was evaluated in rats in a short-term assay in which precancerous lesions were induced by DMH. This animal model is one of the most frequently used for the study of chemopreventive agents since develops morphological and histological features similar to those observed in CRC, which is sporadic and the most common in humans [18]. DMH is an alkylating agent that produce free radicals that bind to DNA and cause mutations [16], specifically in the large intestine upon subcutaneous or intraperitoneal administration in a dose dependent manner [16,18]. The mechanisms postulated to cause these preneoplastic lesions were the accumulation of mutations that drive tumor initiation and then progression [19]. The carcinogen DMH cause mutations in the DNA affecting different genes. Between them, *K-ras* mutations were found to be frequent (between 20–40%) in the first step of CRC process, where ACF are hyperplastic. The mutation of this oncogene encode an intracellular signaling molecule that activate Ras and its down-stream signaling pathways, such as the Raf/MEK/MAPK and PI3K/Akt/PKB [17,20]. In addition, mutations of the β-*catenin* gene have been described in dysplastic MDF, being the activation of Wnt signaling by accumulation of β-*catenin* a major mechanism in the DMH-induced colon carcinogenesis model that leads to cell proliferation [21]. Mutations of *K-ras* and β-*catenin* may be involved in the up-regulation of NF-κB, cyclin D1, COX-2 and iNOS causing an increase in cell proliferation and a decrease of apoptosis [22].

The experimental design used in the present study consisted on three subcutaneous injections of DMH followed by an observation period of 4 weeks, resulted in the formation of preneoplastic lesions, including aberrant crypt foci (ACF) and mucin-depleted foci (MDF) [5]. ACF not only appears in murine animal models induced with carcinogens but also have been described in humans suffering from CRC and FAP [23]. ACF are the first lesions in the development of CRC and show a marked hyperplasia. MDF constitute a more advanced stage than ACF and are characterized by a scarce or absent production of mucins secreted by goblet cells, and exhibit dysplasia [18]. These crypts devoid of mucins have been identified in the colon of humans at high risk of cancer and are considered a hallmark of malignant potential [17]. ACF and MDF are considered as biomarkers of colon carcinogenesis and are used to evaluate the chemopreventive potential of bioactive compounds [24].

Interestingly, the administration of the carcinogen produced preneoplastic lesions in our murine model without any signs of toxicity confirming previous results [5]. When the chemopreventive agent, maslinic acid, was also tested, no adverse effects were observed. Body weight, food conversion efficiency and hepatic enzymes were affected neither by treatment with DMH nor by maslinic acid. These results are consistent with the lack of toxicity reported for this pentacyclic triterpene in different animal species [10,25,26] and humans [27].

In our study, the number of ACF and MDF developed in the positive control rats (DMH+/MA−) was in agreement with the results obtained previously in our group [5] as well as by other authors [18,24,28]. Our findings were also consistent in their distribution along the colon, mostly in the middle and distal segments [18,24,28]. Once the model was established, it was applied to the evaluation of the chemopreventive activity of maslinic acid. Our results indicate that this pentacyclic triterpene exerted a protective activity since ACF in total colon were reduced in a 15%, 18% and 33% at the doses of 5, 10 and 25 mg/kg, respectively. Multiplicity was also affected by maslinic acid, with a significant shortcut in the number of focus with 1, 2, 3, 4 and ≥5 crypts at the dose of 25 mg/kg. MDF were markedly lower in the treated groups, with a reduction of 27% at the lower dose of 5 mg/kg that increased to 51% at 25 mg/kg. Moreover, this compound exerted a halt in the multiplicity of MDF, since the count of focus with 1, 2, 3 and 4 crypts were decreased with respect to the positive control. Taken together our results found in both, ACF and MDF, there is an unequivocal evidence of the chemopreventive potential of maslinic acid in the initiation phase of CRC.

To the best of our knowledge, this is the first report to demonstrate the potential of maslinic acid to reduce DMH-induced ACF and MDF. This murine model of CRC has been widely used to evaluate the chemopreventive activity of other pentacyclic triterpenes such as oleanolic acid, ursolic acid and glycyrrhizic acid, among others [22,29]. In this way, oleanolic acid, that only differs from maslinic acid by the lack of a hydroxyl group at the 2-carbon position, reduced the number of ACF and crypt multiplicity [30,31]. Similar results were found for both oleanolic acid and its positional isomer, ursolic acid [32]. Moreover, glycyrrhizic acid supplementation reduced the number of ACF but not at the significant level. However, MDF were significantly reduced by the treatment [33]. One of the plausible effects of these compounds was the attenuation of the mucin depletion in DMH treated animals due to the anti-inflammatory properties of these pentacyclic triterpenes [22]. In this sense, it has been reported that maslinic acid acts as a suppressor of the pro-inflammatory pathway. This pentacyclic triterpene has been described to inhibit the transcription factor NF-κB which has been involved in the progression of inflammation linked to colon malignancies by down-regulating the expression of iNOs, and COX-2 [34,35]. These key inflammatory molecules have been reported in mucin depleted foci, thus indicating that detectable levels of local inflammation are present in the very early phases of carcinogenesis [23]. Targeting inflammation by maslinic acid provides an important strategy for cancer prevention, since abnormal expression of pro-inflammatory COX-2 have been thought to play a crucial role in colorectal cancer development [22]. Activation of the Wnt/ β-catenin pathway and the inflammation of the colonic mucosa, lead to the activation of transcription factors associated with cell proliferation such as c-Myc, c-Jun and cyclin-D1 as well as to apoptosis such as Bcl-2, Bcl-xl and p53 [22]. The effects exerted by maslinic acid on ACF and more important in the dysplastic MDF indicates a chemopreventive activity on the initiation of colon carcinogenesis. The activity developed in the early phases of carcinogenesis would reinforce those found in colorectal adenocarcinoma cell lines HT-29 and Caco-2 cells. We have previously shown that maslinic acid is capable of inhibiting the growth in HT29 cells and induced apoptosis by the intrinsic pathway, as evidenced by the generation of mitochondrial superoxide anions that served as a pro-apoptotic signal [36]. Subsequently, these results were further confirmed involving the JNK-Bid signaling pathway via the activation of p53 prior to the activation of caspases [12]. However, in Caco-2 cells the pentacyclic triterpene induced apoptosis through a death receptor-mediated apoptotic mechanism [13]. All these data suggest that maslinic acid is able to suppress the activation of multiple CRC pathways to inhibit cell proliferation and to induce apoptosis.

Another issue upon which our study sheds light is on setting up a relationship between the dose of maslinic acid administered and the concentration of this compound reaching the colon. Hence, this knowledge could help establish appropriate doses to produce chemopreventive effects. Our results indicate that the concentrations facing the colonic mucosa 20 h after the last oral administration were around 18, 90 and 450 nmol/g for the doses of 5, 10 and 25 mg/kg, respectively. These results suggest that the doses administered can provide amounts associated with the anti-proliferative and pro-apoptotic activities exerted in HT29 and Caco-2 cells that were within the range of 10 to 250 μM [36, 37]. The link between the amounts of maslinic acid at the target site and the reduction of preneoplastic lesions was assessed by means of the Pearson's correlation coefficient. Worth mentioning the fact that maslinic acid inhibited the formation of the more advanced lesions, represented by MDF, revealing this compound as an interesting molecule for the prevention of CRC. Among the factors related to CRC risk, obesity, physical activity, inflammation and dietary habits have been suggested to play an important role. A diet with a high intake of dietary fiber, vegetables and fruits has been associated not only with a lower incidence of CRC onset [3,4] but also to improve the overall survival in patients with this disease [38]. In the present study, maslinic acid at 5 mg/kg already exerted a chemopreventive activity, and this low dose could be accomplished following a diet rich in foods containing a high amount of this compound. Thus, the intake of one serving size (125 g) of cooked lentils [39], chickpeas [39], eggplant [40], and spinaches [40] will supply 4.9, 7.7, 8.4 and 14.2 mg of maslinic acid, respectively. It is especially remarkable the high content of this triterpene in table olives, since the intake of only 7 units of the Kalamata variety will provide an amount greater than 10 mg of maslinic acid [41]. Therefore, our results not only strengthen the evidence that the intake of maslinic acid protects against CRC, but also could help to develop dietary recommendations of foods rich in this compound for the prevention of this disease.

4. Materials and Methods

4.1. Chemicals and Reagents

Maslinic acid was provided by Dr. Parra from the University of Granada (Granada, Spain). The triterpene was obtained as a pure (>95%) white powder after extraction of the olive pomace with ethyl acetate and further purification by two-step flash chromatography [14]. Betulinic acid, which was used as internal standard (I.S.), was supplied by Extrasynthèse (Genay, France). Ethyl acetate and methanol were from J.T. Baker (Deventer, The Netherlands), whereas acetonitrile was from Scharlau Chemie S.A. (Barcelona, Spain), being all of them LC-MS grade. (2-Hydroxypropyl)-β-cyclodextrin and 10% buffered formalin (pH 7.4) were provided by Sigma-Aldrich S.L. (Tres Cantos, Madrid, Spain). All other chemicals used in the preparation of solutions were of analytical reagent grade. Ultrapure water was obtained by purification through a Milli-Q®Gradient system (Merck Milliore, Madrid, Spain).

4.2. Animals and Diets

Adult male rats (7–8 weeks old) of the Sprague-Dawley strain were obtained from the Animal House Facility of the Facultat de Farmàcia i Ciències de l'Alimentació (Universitat de Barcelona). Animals were housed in cages (n = 2-3/cage) and maintained in a controlled environment with a dark-light cycle of 12 h, relative humidity between 40% and 70% and temperature of 22 ± 2 °C. Rats had free access to both water and food (2014 Teklad Global 14%, Harlan, Barcelona, Spain). Maslinic acid was not detected in the commercial diet analyzed following an extraction method previously described [14]. Animal manipulation was performed in the morning to avoid the effects of circadian rhythms. All experimental procedures met the ethical requirements established by the Guide for the Care and Use of Laboratory Animals and were approved by the Ethics Committee of Animal Experimentation of the Universitat de Barcelona (CEEA-UB ref. 373/12) and the Generalitat de Catalunya (ref. 6558).

4.3. Experimental Design

Rats were randomly distributed into seven groups: Group DMH-/MA- (negative control: no carcinogen, no test agent; $n = 8$), Group DMH–/MA 5 (no carcinogen, 5 mg/kg of maslinic acid; $n = 6$), Group DMH−/MA 10 (no carcinogen, 10 mg/kg of maslinic acid; $n = 6$), Group DMH+/MA− (positive control: DMH, no test agent; $n = 8$), Group DMH+/MA 5 (DMH, 5 mg/kg of maslinic acid; $n = 6$), Group DMH+/MA 10 (DMH, 10 mg/kg of maslinic acid; $n = 6$) and Group DMH+/MA 25 (DMH, 25 mg/kg of maslinic acid; $n = 6$). Maslinic acid was administered daily by oral gavage (10 mL/kg) at doses of 5, 10 or 25 mg/kg over 49 days. Due to its low solubility in water, an aqueous solution of the triterpene was obtained by means of (2-hydroxypropyl)-β-cyclodextrin 20% (w/v) [10]. This solution was prepared weekly and kept at 4 °C. Rats in the groups DMH−/MA− and DMH+/MA− received only the solvent. On days 8, 15 and 22 of the experimental period, the carcinogenic agent dissolved in EDTA 1 mmol/L (pH 6.5) was administered at a dose of 20 mg/kg by intraperitoneal injections (1 mL/kg). DMH doses were freshly prepared before each use. Animals in the three DMH– groups were given an intraperitoneal injection of EDTA 1 mmol/L.

Body weight was recorded daily, and food and water consumption were monitored every two days throughout the study. Food conversion efficiency (FCE) was calculated, as a percentage, dividing the weekly body weight gain by the weekly food consumption.

4.4. Sample Collection

At the end of the experimental period, overnight fasted rats were anesthetized with ketamine and xylacine (90 and 10 mg/kg, respectively), the abdomen was opened by a midline longitudinal incision and the colon of each animal was resected. The intestinal lumen was rinsed with 5 mL of ice-cold phosphate-buffered solution (PBS) to collect the intestinal content for the quantification of maslinic acid. These samples were rapidly immersed in liquid N_2 and stored at -20 °C until analysis. Subsequently, colons were trimmed of mesenteric fat, cut open on the median axis and divided into three segments of similar length: proximal (close to the caecum), medial and distal (close to the rectum). The length, width and wet weight of each segment were recorded before being fixed flat onto a polystyrene board and plunged in 10% buffered formalin pH 7,4 (Sigma Aldrich, S.L.) for at least 24 h.

4.5. Aberrant Crypt Foci

Once fixed, colon segments were stained with histological dyes for the identification and count of preneoplastic lesions, namely aberrant crypt foci (ACF). For the evaluation of ACF, a modification of a method previously described was used [42]. Briefly, colon segments were dyed with methylene blue 0.2% for 8 min (proximal) or 10 min (medial and distal). The excess of dye was removed by rinsing the tissues with PBS and each segment was placed with the mucosal side up on a microscopic slide prior to examination at the light microscope at magnification of ×10 (Leitz, Leica Microsistemas S.L.U., Barcelona, Spain). Lesions were characterized by a larger size (2–3 times that of normal surrounding crypts), and by displaying a more intense stain, distortion of the opening of the lumen and elevation above the surface of the mucosa [42]. For each animal, the number of foci of aberrant crypts (ACF) were determined in the proximal, medial and distal segments, and were also expressed for the entire colon. Multiplicity was assessed by counting the number of aberrant crypts (AC) forming each focus (AC/ACF). Finally, the total AC in the entire colon were also calculated. The lesions were determined by two independent observers who were blinded to the treatments.

4.6. Mucin Depleted Foci

Following ACF count, colon segments were kept in PBS at 4 °C until being processed with the high-iron diamine/Alcian blue/neutral red staining (HID-AB) for the observation of mucin production. Tissues were rinsed with PBS prior to being immersed in the high-iron diamine solution for 18–24 h protected from the light. Then, segments were washed in PBS prior to being stained with alcian blue 1% in acetic acid 3% for 5 min, then rinsed again, and stained for 2 min in neutral red 0.1% in acetic acid 0.002%. Tissues were mounted on microscopic slides and examined under a light microscope at magnification of ×20. The total number of mucin depleted foci (MDF), multiplicity expressed as mucin-depleted aberrant crypts per aberrant focus (MDAC/MDF) and total mucin-depleted aberrant crypts (MDAC) were assessed in the HID-AB stained colons [43]. MDF are characterized by absence or little production of mucins, distortion of the opening of the lumen as compared with normal crypts, and elevation of the lesion above the surface of the mucosa. As stated above, the scores were evaluated by two independent observers blinded to the groups.

4.7. Determination of AST and ALT

Aspartate aminotransferase (AST) and alanine aminotransferase (ALT) were determined in the following three representative groups: DMH−/MA− (no carcinogen, no test agent; $n = 8$), DMH+/MA− (DMH, no test agent; $n = 8$) and DMH+/MA 25 (DMH, 25 mg/kg of maslinic acid; $n = 7$). Hence, blood was collected from anesthetized rats by cardiac puncture and was transferred into a tube without anticoagulant for the determination of hepatic enzymes. Serum was obtained after centrifugation of blood samples at 1500× g (Megafuge 1.0R, Heraeus, Boadilla, Spain) for 15 min at 4 °C. Analyses of serum were carried out with a Roche/Hitachi 747 clinical analyzer from Roche Diagnostics GmbH (Mannheim, Germany).

4.8. Determination of Maslinic Acid in Colon Content

The concentration of maslinic acid in colon content was determined by liquid extraction prior to LC-APCI-MS analysis as previously described [14]. The quantification was carried out in the groups that were induced preneoplastic lesions and received maslinic acid at 0, 5, 10 and 25 mg/kg. To assess whether DMH treatment could affect the oral bioavailability of maslinic acid, the colon contents of the rats in the group that were not injected with the carcinogen but received 10 mg/kg of maslinic acid were also analyzed. In all the groups, samples were obtained between 18–20 h after the last oral administration.

4.9. Statistical Analysis

Results are presented as means ± standard error of the mean (SEM). GraphPad Prism 6 (GraphPad Software, Inc., La Joya, CA, USA) was used for data evaluation and statistical analyses. Normality of the data was evaluated by the Kolmogorov-Smirnov test, and depending on the significance a parametric or non-parametric analysis was applied. Body weight and food conversion efficiency were compared using a two-way analysis of variance (ANOVA) followed by Tukey's multiple comparisons test. The number of ACF, MDF in total colon, and segments as well as multiplicity of ACF and MDF were analyzed by the non-parametric Kruskal–Wallis test, followed by Dunn's multiple comparisons test. AST and ALT in serum, as well as the concentration of maslinic acid in colon content were evaluated with one-way ANOVA and Bonferroni's multiple comparisons test. The correlations between the percentage of reduction of ACF and MDF colonic content of maslinic acid were assessed by Pearson's correlation method, previous application of Kolmogorv-Smirnov test. For all tests, two levels of significance were considered, $p < 0.05$ and $p < 0.01$.

5. Conclusions

The results described here demonstrate that the daily ingestion of maslinic acid at doses as low as 5 mg/kg reduces the formation of ACF and most importantly the dysplastic MDF lesions in a rat model relevant to human colorectal carcinogenesis. Consequently, and based on the results, it is not unreasonable to associate maslinic acid to the lower prevalence of CRC linked to a greater adherence to a Mediterranean diet [1,2]. Therefore, this pentacyclic triterpene merits further clinical evaluation in CRC chemoprevention.

Author Contributions: Conceptualization, J.M.P.; Data curation, G.L.-M. and M.S.-G.; Formal analysis, M.E.J., G.L.-M. and J.M.P.; Funding acquisition, J.M.P.; Investigation, G.L.-M. and M.S.-G.; Methodology, M.E.J. and J.M.P.; Project administration, J.M.P.; Resources, J.M.P.; Supervision, M.E.J. and J.M.P.; Validation, M.E.J. and J.M.P.; Visualization, M.E.J., G.L.-M. and J.M.P.; Writing—original draft, M.E.J., G.L.-M. and J.M.P.; Writing—review & editing, M.E.J. and J.M.P.

Funding: This research was funded by grant number AGL2009-12866 from Ministerio de Ciencia e Innovación, grant number AGL2013-41188 from Ministerio de Economía y Competividad and grants number 2009SGR471, 2014SGR1221 and 2017SGR945 from Generalitat de Catalunya. G.L.-M. was the recipient of the training grant Ajuts de Personal Investigador en Formació (APIF) de la Universitat de Barcelona. M.S.-G. was funded by project 2009SGR471.

Conflicts of Interest: The authors declare no conflicts of interest.

References

1. Ferlay, J.; Colombet, M.; Soerjomataram, I.; Mathers, C.; Parkin, D.M.; Piñeros, M.; Znaor, A.; Bray, F. Estimating the global cancer incidence and mortality in 2018: GLOBOCAN sources and methods. *Int. J. Cancer* **2018**, *00*, 1–13. [CrossRef] [PubMed]
2. Farinetti, A.; Zurlo, V.; Manenti, A.; Coppi, F.; Mattioli, A.V. Mediterranean diet and colorectal cancer: A systematic review. *Nutrition* **2017**, *43*, 83–88. [CrossRef]
3. Haslam, A.; Robb, S.W.; Hébert, J.R.; Huang, H.; Ebell, M.H. Greater adherence to a Mediterranean diet is associated with lower prevalence of colorectal adenomas in men of all races. *Nutr. Res.* **2017**, *48*, 76–84. [CrossRef] [PubMed]
4. Park, S.Y.; Boushey, C.J.; Wilkens, L.R.; Haiman, C.A.; Le Marchand, L. High-quality diets associate with reduced risk of colorectal cancer: Analyses of diet quality indexes in the multiethnic cohort. *Gastroenterology* **2017**, *153*, 386–394. [CrossRef] [PubMed]
5. Alfaras, I.; Juan, M.E.; Planas, J.M. *trans*-Resveratrol reduces precancerous colonic lesions in dimethylhydrazine-treated rats. *J. Agric. Food Chem.* **2010**, *58*, 8104–8110. [CrossRef] [PubMed]
6. Hu, R.; Khor, T.O.; Shen, G.; Jeong, W.S.; Hebbar, V.; Chen, C.; Xu, C.; Reddy, B.; Chada, K.; Kong, A.N. Cancer chemoprevention of intestinal polyposis in $Apc^{Min/+}$ mice by sulforaphane, a natural product derived from cruciferous vegetable. *Carcinogenesis* **2006**, *27*, 2038–2046. [CrossRef] [PubMed]
7. Pandurangan, A.K.; Kumar, S.A.; Dharmalingam, P.; Ganapasam, S. Luteolin, a bioflavonoid inhibits azoxymethane-induced colon carcinogenesis: Involvement of iNOS and COX-2. *Pharmacogn. Mag.* **2014**, *10*, S306–S310. [PubMed]
8. Neergheen, V.S.; Bahorun, T.; Taylor, E.W.; Jen, L.S.; Aruoma, O.I. Targeting specific cell signaling transduction pathways by dietary and medicinal phytochemicals in cancer chemoprevention. *Toxicology* **2010**, *278*, 229–241. [CrossRef] [PubMed]
9. Lozano-Mena, G.; Sánchez-González, M.; Juan, M.E.; Planas, J.M. Maslinic acid, a natural phytoalexin-type triterpene from olives-a promising nutraceutical? *Molecules* **2014**, *19*, 11538–11559. [CrossRef] [PubMed]
10. Sánchez-González, M.; Lozano-Mena, G.; Juan, M.E.; García-Granados, A.; Planas, J.M. Assessment of the safety of maslinic acid, a bioactive compound from *Olea europaea* L. *Mol. Nutr. Food Res.* **2013**, *57*, 339–346. [CrossRef]
11. Juan, M.E.; Wenzel, U.; Ruiz-Gutierrez, V.; Daniel, H.; Planas, J.M. Olive fruit extracts inhibit proliferation and induce apoptosis in HT-29 human colon cancer cells. *J. Nutr.* **2006**, *136*, 2553–2557. [CrossRef] [PubMed]
12. Reyes-Zurita, F.J.; Pachón-Peña, G.; Lizárraga, D.; Rufino-Palomares, E.E.; Cascante, M.; Lupiáñez, J.A. The natural triterpene maslinic acid induces apoptosis in HT29 colon cancer cells by a JNK-p53-dependent mechanism. *BMC Cancer* **2011**, *11*, 154–167. [CrossRef] [PubMed]

13. Reyes-Zurita, F.J.; Rufino-Palomares, E.E.; García-Salguero, L.; Peragón, J.; Medina, P.P.; Parra, A.; Cascante, M.; Lupiáñez, J.A. Maslinic acid, a natural triterpene, induces a death receptor-mediated apoptotic mechanism in Caco-2 p53-deficient colon adenocarcinoma cells. *PloS ONE* **2016**, *11*, e0146178–e0146194. [CrossRef] [PubMed]
14. Lozano-Mena, G.; Sánchez-González, M.; Parra, A.; Juan, M.E.; Planas, J.M. Identification of gut-derived metabolites of maslinic acid, a bioactive compound from *Olea europaea* L. *Mol. Nutr. Food Res.* **2016**, *60*, 2053–2064. [CrossRef]
15. Sánchez-Tena, S.; Reyes-Zurita, F.J.; Díaz-Moralli, S.; Vinardell, M.P.; Reed, M.; García-Garía, F.; Dopazo, J.; Lupiáñez, J.A.; Günther, U.; Cascante, M. Maslinic acid-enriched diet decreases intestinal tumorigenesis in $Apc^{Min/+}$ mice through transcriptomic and metabolomic reprogramming. *PLoS ONE* **2013**, *8*, e59392–e59403.
16. Rosenberg, D.W.; Giardina, C.; Tanaka, T. Mouse models for the study of colon carcinogenesis. *Carcinogenesis* **2009**, *30*, 183–196. [CrossRef] [PubMed]
17. Femia, A.P.; Tarquini, E.; Salvadori, M.; Ferri, S.; Giannini, A.; Dolara, P.; Caderni, G. K-ras mutations and mucin profile in preneoplastic lesions and colon tumors induced in rats by 1,2-dimethylhydrazine. *Int. J. Cancer* **2008**, *122*, 117–123.
18. Perše, M.; Cerar, A. Morphological and molecular alterations in 1,2 dimethylhydrazine and azoxymethane induced colon carcinogenesis in rats. *J. Biomed. Biotechnol.* **2011**, *2011*, 473964–473978. [CrossRef]
19. Fearon, E.R.; Vogelstein, B. A genetic model for colorectal tumorigenesis. *Cell* **1990**, *61*, 759–767. [CrossRef]
20. Takahashi, M.; Wakabayashi, K. Gene mutations and altered gene expression in azoxymethane-induced colon carcinogenesis in rodents. *Cancer Sci.* **2004**, *95*, 475–480. [CrossRef]
21. Femia, A.P.; Dolara, P.; Giannini, A.; Salvadori, M.; Biggeri, A.; Caderni, G. Frequent mutation of *Apc* gene in rat colon tumors and mucin-depleted foci, preneoplastic lesions in experimental colon carcinogenesis. *Cancer Res.* **2007**, *67*, 445–449. [CrossRef] [PubMed]
22. Sharma, S.H.; Thulasingam, S.; Nagarajan, S. Terpenoids as anti-colon cancer agents—A comprehensive review on its mechanistic perspectives. *Eur. J. Pharmacol.* **2017**, *795*, 169–178. [CrossRef] [PubMed]
23. Femia, A.P.; Dolara, P.; Luceri, C.; Salvadori, M.; Caderni, G. Mucin-depleted foci show strong activation of inflammatory markers in 1,2-dimethylhydrazine-induced carcinogenesis and are promoted by the inflammatory agent sodium dextran sulfate. *Int. J. Cancer* **2009**, *125*, 541–547. [CrossRef]
24. Raju, J. Azoxymethane-induced rat aberrant crypt foci: relevance in studying chemoprevention of colon cancer. *World J. Gastroenterol.* **2008**, *21*, 6632–6635. [CrossRef]
25. Yan, S.L.; Yang, H.T.; Lee, H.L.; Yin, M.C. Protective effects of maslinic acid against alcohol-induced acute liver injury in mice. *Food Chem. Toxicol.* **2014**, *74*, 149–155. [CrossRef]
26. Hidalgo, M.C.; Skalli, A.; Abellán, E.; Arizcun, M.; Cardenete, G. Dietary intake of probiotics and maslinic acid in juvenile dentex (*Dentex dentex* L.): Effects on growth performance, survival and liver proteolytic activities. *Aquaculture Nutr.* **2006**, *12*, 256–266. [CrossRef]
27. Fukumitsu, S.; Villareal, M.O.; Aida, K.; Hino, A.; Hori, N.; Isoda, H.; Naito, Y. Maslinic acid in olive fruit alleviates mild knee joint pain and improves quality of life by promoting weight loss in the elderly. *J. Clin. Biochem. Nutr.* **2016**, *59*, 220–225. [CrossRef]
28. Bird, R.P.; Good, C.K. The significance of aberrant crypt foci in understanding the pathogenesis of colon cancer. *Toxicol. Lett.* **2000**, *112–113*, 395–402. [CrossRef]
29. Zhao, Y.; Hu, X.; Zuo, X.; Wang, M. Chemopreventive effects of some popular phytochemicals on human colon cancer: A review. *Food Funct.* **2018**, *9*, 4548–4568. [CrossRef] [PubMed]
30. Kawamori, T.; Tanaka, T.; Hara, A.; Yamahara, J.; Mori, H. Modifying effects of naturally occurring products on the development of colonic aberrant crypt foci induced by azoxymethane in F344 rats. *Cancer Res.* **1995**, *55*, 1277–1282. [PubMed]
31. Janakiram, N.B.; Indranie, C.; Malisetty, S.V.; Jagan, P.; Steele, V.E.; Rao, C.V. Chemoprevention of colon carcinogenesis by oleanolic acid and its analog in male F344 rats and modulation of COX-2 and apoptosis in human colon HT-29 cancer cells. *Pharm. Res.* **2008**, *25*, 2151–2157. [CrossRef] [PubMed]
32. Furtado, R.A.; Rodrigues, E.P.; Araújo, F.R.; Oliveira, W.L.; Furtado, M.A.; Castro, M.B.; Cunha, W.R.; Tavares, D.C. Ursolic acid and oleanolic acid suppress preneoplastic lesions induced by 1,2-dimethylhydrazine in rat colon. *Toxicol Pathol.* **2008**, *36*, 576–580. [CrossRef] [PubMed]

33. Khan, R.; Khan, A.Q.; Lateef, A.; Rehman, M.U.; Tahir, M.; Ali, F.; Hamiza, O.O.; Sultana, S. Glycyrrhizic acid suppresses the development of precancerous lesions via regulating the hyperproliferation, inflammation, angiogenesis and apoptosis in the colon of Wistar rats. *PloS ONE* **2013**, *8*, e56020–e56042. [CrossRef] [PubMed]
34. Li, C.; Yang, Z.; Zhai, C.; Qiu, W.; Li, D.; Yi, Z.; Wang, L.; Tang, J.; Qian, M.; Luo, J.; Liu, M. Maslinic acid potentiates the anti-tumor activity of tumor necrosis factor alpha by inhibiting NF-kappaB signaling pathway. *Mol. Cancer* **2010**, *9*, 73–86. [CrossRef]
35. Hsum, Y.W.; Yew, W.T.; Hong, P.L.; Soo, K.K.; Hoon, L.S.; Chieng, Y.C.; Mooi, L.Y. Cancer chemopreventive activity of maslinic acid: Suppression of COX-2 expression and inhibition of NF-κB and AP-1 activation in Raji cells. *Planta Med.* **2011**, *77*, 152–157. [CrossRef]
36. Juan, M.E.; Planas, J.M.; Ruiz-Gutiérrez, V.; Daniel, H.; Wenzel, U. Antiproliferative and apoptosis-inducing effects of maslinic and oleanolic acids, two pentacyclic triterpenes from olives, on HT-29 colon cancer cells. *Br. J. Nutr.* **2008**, *100*, 36–43. [CrossRef]
37. Reyes-Zurita, F.J.; Rufino-Palomares, E.E.; Medina, P.P.; García-Salguero, E.L.; Peragón, J.; Cascante, M.; Lupiáñez, J.A. Antitumour activity on extrinsic apoptotic targets of the triterpenoid maslinic acid in p53-deficient Caco-2 adenocarcinoma cells. *Biochimie* **2013**, *95*, 2157–2167. [CrossRef]
38. Ratjen, I.; Schafmayer, C.; di Giuseppe, R.; Waniek, S.; Plachta-Danielzik, S.; Koch, M.; Nöthlings, U.; Hampe, J.; Schlesinger, S.; Lieb, W. Postdiagnostic mediterranean and healthy nordic dietary patterns are inversely associated with all-cause mortality in long-term colorectal cancer survivors. *J. Nutr.* **2017**, *147*, 636–644. [CrossRef]
39. Kalogeropoulos, N.; Chiou, A.; Ioannou, M.; Karathanos, V.T.; Hassapidou, M.; Andrikopoulos, N.K. Nutritional evaluation and bioactive microconstituents (phytosterols, tocopherols, polyphenols, triterpenic acids) in cooked dry legumes usually consumed in the Mediterranean countries. *Food Chem.* **2010**, *121*, 682–690. [CrossRef]
40. Lin, C.C.; Huang, C.Y.; Mong, M.C.; Chan, C.Y.; Yin, M.C. Antiangiogenic potential of three triterpenic acids in human liver cancer cells. *J. Agric. Food Chem.* **2011**, *59*, 755–762. [CrossRef] [PubMed]
41. Romero, C.; García, A.; Medina, E.; Ruíz-Méndez, M.A.; de Castro, A.; Brenes, M. Triterpenic acids in table olives. *Food Chem.* **2010**, *118*, 670–674. [CrossRef]
42. Bird, R.P. Observation and quantification of aberrant crypts in the murine colon treated with a colon carcinogen: preliminary findings. *Cancer Lett.* **1987**, *37*, 147–151. [CrossRef]
43. Caderni, G.; Femia, A.P.; Giannini, A.; Favuzza, A.; Luceri, C.; Salvadori, M.; Dolara, P. Identification of mucin-depleted foci in the unsectioned colon of azoxymethane-treated rats: correlation with carcinogenesis. *Cancer Res.* **2003**, *63*, 2388–2392. [PubMed]

molecules

MDPI

Article

Untargeted Metabolomics Study of the In Vitro Anti-Hepatoma Effect of Saikosaponin d in Combination with NRP-1 Knockdown

Yingtong Lv [1,2], Xiaoying Hou [1,2], Qianqian Zhang [1,2], Ruiting Li [1,2], Lei Xu [1,2], Yadong Chen [3], Yuan Tian [1,2], Rong Sun [4], Zunjian Zhang [1,2,*] and Fengguo Xu [1,2,*]

1 Key Laboratory of Drug Quality Control and Pharmacovigilance (Ministry of Education), State Key Laboratory of Natural Medicine, China Pharmaceutical University, Nanjing 210009, China; lv_yingtong@126.com (Y.L.); cpuhxy_2011@163.com (X.H.); zhangqianqian09413@163.com (Q.Z.); 15051853218@163.com (R.L.); 18356002797@163.com (L.X.); 1020041334@cpu.edu.cn (Y.T.)

2 Jiangsu Key Laboratory of Drug Screening, China Pharmaceutical University, Nanjing 210009, China

3 Department of Organic Chemistry, China Pharmaceutical University, Nanjing 210009, China; ydchen@cpu.edu.cn

4 Advanced Medical Research Institute, Shandong University, Jinan 250100, China; sunrong107@163.com

* Correspondence: zzj@cpu.edu.cn (Z.Z.); fengguoxu@cpu.edu.cn (F.X.); Tel.: +86-25-83271021 (F.X.)

Received: 1 March 2019; Accepted: 10 April 2019; Published: 11 April 2019

check for updates

Abstract: Saikosaponin d (SSd) is one of the main active ingredients in Radix Bupleuri. In our study, network pharmacology databases and metabolomics were used in combination to explore the new targets and reveal the in-depth mechanism of SSd. A total of 35 potential targets were chosen through database searching (HIT and TCMID), literature mining, or chemical similarity predicting (Pubchem). Out of these obtained targets, Neuropilin-1 (NRP-1) was selected for further research based on the degree of molecular docking scores and novelty. Cell viability and wound healing assays demonstrated that SSd combined with NRP-1 knockdown could significantly enhance the damage of HepG2. Metabolomics analysis was then performed to explore the underlying mechanism. The overall difference between groups was quantitatively evaluated by the metabolite deregulation score (MDS). Results showed that NRP-1 knockdown exhibited the lowest MDS, which demonstrated that the metabolic profile experienced the slightest interference. However, SSd alone, or NRP-1 knockdown in combination with SSd, were both significantly influenced. Differential metabolites mainly involved short- or long-chain carnitines and phospholipids. Further metabolic pathway analysis revealed that disturbed lipid transportation and phospholipid metabolism probably contributed to the enhanced anti-hepatoma effect by NRP-1 knockdown in combination with SSd. Taken together, in this study, we provided possible interaction mechanisms between SSd and its predicted target NRP-1.

Keywords: saikosaponin d; neuropilin-1; HepG2; metabolomics; metabolite deregulation score

1. Introduction

Radix Bupleuri is a common traditional Chinese medicine (TCM), which has been used in China for over 2000 years. Saikosaponin d (SSd), as one of the main active ingredients extracted from Radix Bupleuri, has been proved to be liver-protective and is used to treat liver fibrogenesis [1], inflammation [2], and viral hepatitis. It was also reported that SSd exerted antitumor [3], immunoregulation [4], neuroregulation [5], and anti-allergic activities [6]. However, on the contrary, studies showed that SSd also had toxicity and could induce liver injury [7], hepatocyte or hepatic stellate cells apoptosis [8,9]. The effect of SSd is diverse and the underlying mechanisms remain unclear.

Target predicting is a feasible strategy that could facilitate a mechanistic study of TCM. For TCM, databases such as HIT, TCMSP, TCM Database@Taiwan, and TCMID can be used to search for potential targets [10–12]. However, databases cannot update reported targets. On the other hand, different databases contain different targets due to various algorithms. There is no database integrating targets from all sources yet. To further explore unreported targets, predicting approaches based on chemoinformatics could be applied, such as ligand-based prediction, receptor-based prediction, and data mining-based prediction [13–16]. However, the three-dimensional structures of targets are usually ignored, so the mutual binding mode of drug and targets cannot be fully reflected. Thus, for the targets obtained, molecular docking can be applied to evaluate the binding affinity between SSd and proteins, which is a fast, accurate method to predict structural affinity.

NRP-1 is a transmembrane glycoprotein composed of five domains (a1, a2, b1, b2, c) [17], whereas a1/a2 subunit mainly binds class-3 semaphorins and transduces signals by conjunction with plexins [18]. Vascular endothelial growth factors (VEGF) interact with b1/b2 subunit to exert angiogenesis regulation effects [19]. Studies have shown that NRP-1 was upregulated in liver cancer [20], prostate tumor [21], gastric cancer [22], breast cancer [23], lung carcinomas, etc. Overexpression of NRP-1 was found to promote tumor angiogenesis as well as the growth and proliferation of cancer cells. On the other hand, NRP-1 plays an important role in the immune system. NRP-1 expression in immune cells such as tumor-associated macrophages, T lymphocytes, and dendritic cells is related to tumor growth, migration, and cell-cell interaction [24,25].

Metabolomics is a powerful approach that could reflect dynamic changes of small molecule metabolites when disturbance such as disease and drug-taking happens [26]. Since metabolites are located at the end of a biological information flow, changes could reflect the signal amplification effects of biochemical changes. Therefore, metabolomics could act as a tool to reveal the mechanism between TCM and potential targets.

In our present study, in order to explore new targets and reveal the in-depth mechanism of SSd, network pharmacology and metabolomics were used in combination. First, potential targets of SSd were obtained by database searching, literature mining, or chemical similarity predicting. Then molecular docking was performed to investigate the binding affinity of SSd with potential targets. Finally, untargeted metabolomics analysis was carried out to reveal the possible mechanism by which SSd affected the aimed target and the role of NRP-1.

2. Results

2.1. NRP-1 Was Selected as a Target of SSd

Fourteen targets of SSd were obtained from the HIT [10] and TCMID [11] databases (Table S1). Furthermore, 19 proteins were collected by literature mining and 14 targets were filtered from Pubchem by chemical similarity predicting (Tables S1–S3) [27]. Targets obtained by the three methods overlapped in some cases, which proved the reliability of the methods, as shown in Figure S1. A total of 35 proteins were pooled as a target set (Table S4). Then SSd was docked with all the targets. Twenty proteins were found to bind with SSd and were ranked by docking scores (Table 1). Prostaglandin G/H synthase 2 (COX-2) showed the strongest interaction with SSd, followed by neuropilin-1 (NRP-1) and transcription factor p65. COX-2 was reported to inhibit DEN-induced liver cancer in rats by SSd [28]. Others found that SSd downregulated transcription factor p65 and mediated inflammatory signaling to protect mice from hepatotoxicity induced by acetaminophen (APAP) [29]. However, the interaction between SSd and NRP-1 was not reported. Thus, we selected NRP-1 for further study.

Table 1. Docking results of SSd and potential targets.

Number	PDB ID	Gene Name	Uniprot ID	Target Name	Docking Score	Validated or Not
1	5F19	PTGS2	P35354	Prostaglandin G/H synthase 2	−6.67	Validated [28]
2	2QQI	NRP1	O14786	Neuropilin−1	−6.469	Unvalidated
3	1NFI	RELA	Q04206	Transcription factor p65	−5.585	Validated [1]
4	2W96	CDK4	P11802	Cell division protein kinase 4	−5.577	Validated [30]
5	1IKN	NFKBIA	P25963	NF-kappa-B inhibitor alpha	−5.346	Validated [1]
6	5JFD	F2	P00734	Prothrombin	−5.325	Unvalidated
7	4AGN	P53	P04637	Cellular tumor antigen p53	−5.205	Validated [31]
8	4MAN	BCL-2	P10415	Apoptosis regulator Bcl-2	−4.857	Validated [7]
9	5TBE	MAPK14	Q16539	Mitogen-activated protein kinase 14	−4.792	Validated [32]
10	1IL6	IL6	P05231	Interleukin-6	−4.666	Validated [1]
11	3V3K	CASP9	P55211	Caspase-9	−4.49	Validated [8]
12	4PRY	CASP3	P42574	Caspase-3	−4.443	Validated [8]
13	5T46	EIF4G1	Q04637	Eukaryotic translation initiation factor 4	−4.414	Unvalidated
14	4LXO	FN1	P02751	Fibronectin	−4.303	Unvalidated
15	2DBF	NFKB1	P19838	Nuclear factor NF-kappa-B p105	−3.902	Unvalidated
16	5FF0	TGFB1	P01137	Transforming growth factor beta-1	−3.642	Validated [33]
17	5T01	JUN	P05412	Transcription factor AP-1	−3.611	Validated [32]
18	4S0O	BAX	Q07812	Apoptosis regulator BAX	−3.412	Validated [8]
19	4FDL	CASP7	P55210	Caspase-7	−3.328	Validated [34]
20	5I4Z	MYC	P01106	Myc proto-oncogene protein	−2.997	Validated [35]

2.2. NRP-1 Knockdown Enhanced the Anti-Hepatoma Effect of SSd

NRP-1 has been reported to display higher expression in hepatocellular carcinoma cell lines than in normal hepatic cell lines [36]. In addition, the published data indicate that a tumor was suppressed by silencing NRP-1. Therefore, it was speculated that the anti-hepatoma effect of SSd might be enhanced if we could downregulate NRP-1 [37,38]. In order to select the most suitable cell lines, we compared NRP-1 expression levels in hepatocellular carcinoma cell lines HepG2 and SMMC-7721 and a normal hepatic cell line L-02 by Western blot (Figure 1a). High NRP-1 expression was detected in the HepG2 cell line, followed by SMMC-7721. L-02 showed the lowest expression of NRP-1, as reported [38]. Consequently, HepG2 was chosen for further study.

Firstly, different concentrations of SSd were used to treat HepG2 and the IC_{50} was 16.02 ± 0.91 μM. The effect of SSd on migration of HepG2 was also measured, as shown in Figure 1c. The results demonstrated that SSd could inhibit the migration of HepG2. Next we investigated NRP-1 levels influenced by SSd. Upregulation of NRP-1 was observed when treated with 3.75 μM and 7.5 μM SSd. However, NRP-1 expression displayed no significant difference at 15 μM compared to the control, as shown in Figure 2a. According to the cell viability, HepG2 was damaged more severely at 15 μM SSd than at 7.5 μM or 3.75 μM, which suggested the mechanism at 15 μM or higher may not be directly related to NRP-1. So we selected a concentration of SSd at 7.5 μM for further investigation.

After NRP-1 was knocked down, HepG2 was treated with SSd. Then both cell viability and migration were measured. Compared to the siNRP-1 (small interfering RNA of gene NRP-1) and SSd-treated NC (negative control) groups, cell viability and migration were decreased significantly in the SSd-treated siNRP-1 group, as shown in Figure 2b–d. The results suggested that NRP-1 knockdown could significantly enhance the anti-hepatoma effect of SSd.

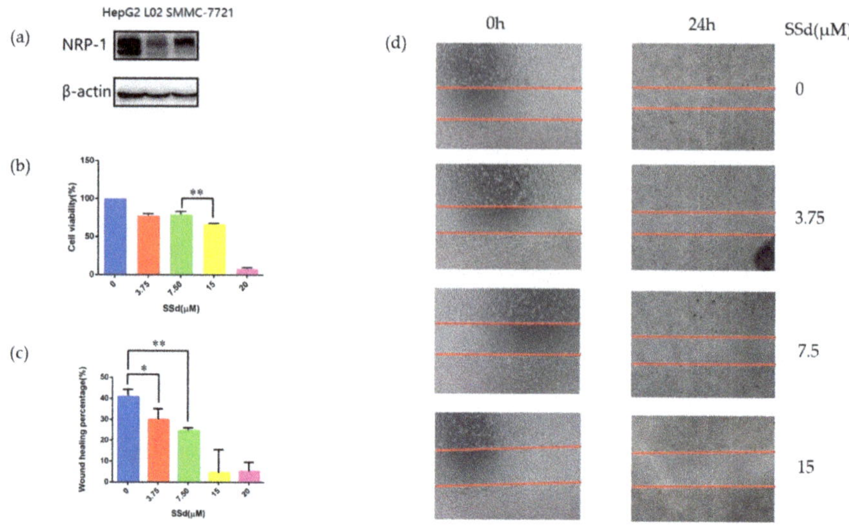

Figure 1. Cell lines comparison and the effects of SSd on HepG2. (**a**) Expression of NRP-1 in three cell lines; (**b**) cell viability of HepG2 after treatment with SSd, ** $p < 0.01$; (**c**) wound healing percentage of HepG2 after treatment with SSd, * $p < 0.05$, ** $p < 0.01$; (**d**) wound healing pictures of HepG2 after treatment with SSd.

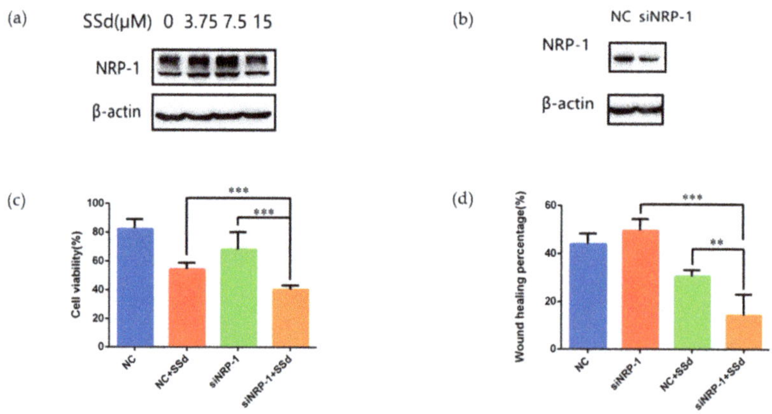

Figure 2. NRP-1 knockdown and effects of SSd on HepG2 in combination with NRP-1 knockdown. (**a**) The expression of NRP-1 affected by SSd; (**b**) knockdown of NRP-1; (**c**) the cell viability of HepG2 cells after knockdown of NRP-1 in combination with SSd, *** $p < 0.001$; (**d**) the wound healing percentage of HepG2 cells after knockdown of NRP-1 in combination with SSd, ** $p < 0.01$, *** $p < 0.001$.

2.3. Quantitative Evaluation of the Effect of NRP-1 Knockdown and SSd by Metabolomics

For all the metabolites obtained by GC/MS and LC/MS, PCA and OPLS-DA models were constructed, as shown in Figure 3. Results from GC/MS and LC/MS were all integrated into one PCA or OPLS-DA plot. In the PCA models, all the QC samples were clustered closely, which indicated a satisfactory analytical performance. The supervised OPLS-DA models showed clearly separation among the four groups and model statistics, R^2X, R^2Y and Q^2 suggested the robustness of the models (Table S5).

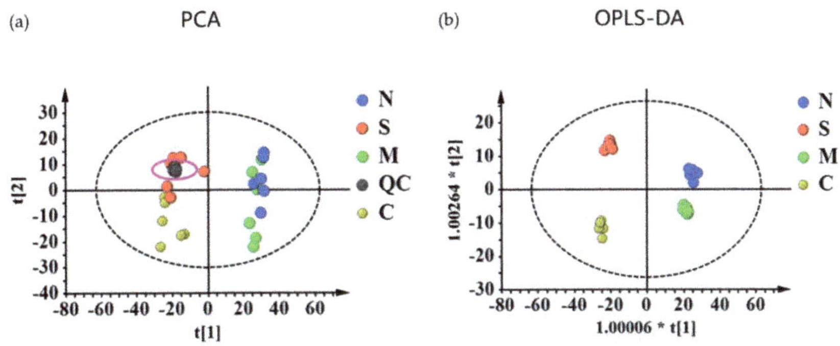

Figure 3. (**a**) Score plots of principal components analysis (PCA) and (**b**) orthogonal partial least squares discriminant analysis (OPLS-DA) models. C:NC samples, M:NC samples treated with SSd, S: siNRP-1 samples, N: siNRP-1 samples treated with SSd.

To further evaluate the influence of NRP-1 and SSd quantitatively, MDS was calculated. The results (Figure 4b) showed that the NRP-1 knockdown group displayed a smaller distance from the principal curve than the combination with SSd or SSd alone, which showed that the metabolic profile had a minor influence. On the other hand, a strong disturbance of the metabolic profile was demonstrated in the SSd-treated NC group as well as the SSd-treated siNRP-1 group.

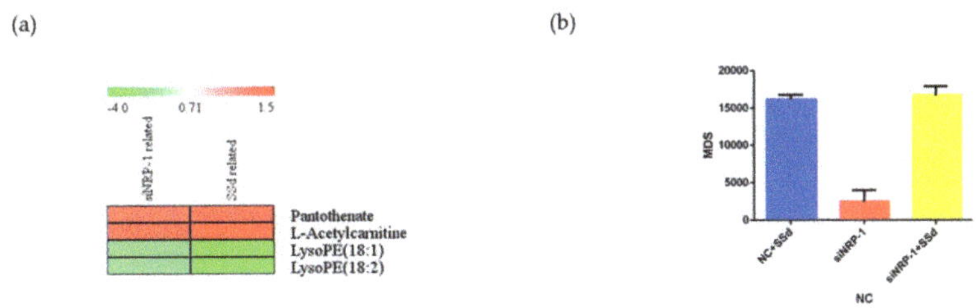

Figure 4. Fold changes of differential metabolites and MDS of all metabolites. (**a**) Heatmap of the fold changes of differential metabolites related to NRP-1 knockdown and SSd-treated; (**b**) MDS of all the metabolites detected with NC as control.

2.4. Carnitines and Phospholipids Were Focalized Based on Metabolomics

Variables with VIP > 1 and $p < 0.05$ were considered differential metabolites. For LC/MS, mass-to-charge (m/z) and MS/MS fragmentation patterns of metabolites were compared to the information provided in the HMDB database for annotation. Metabolites obtained by GC/MS were annotated by comparison with the National Institute of Standard and Technology library. Six differential metabolites were related to NRP-1 knockdown, 30 related to SSd-treated, and four appeared relevant to both. In general, NRP-1 knockdown combined with SSd treatment affected lyso-phosphatidylethanolamines, acetylcarnitine, and pantothenate. For those only affected by one factor, propionylcarnitine and meso-erythritol were related to NRP-1 knockdown, while hexadecenoylcarnitine, dodecanoic acid, galactose, 1-monooleoylglycerol, lyso-phosphatidylcholines, lyso-phosphatidylethanolamines, phosphatidylcholines and phosphatidylethanolamines were disturbed by SSd treatment. The fold changes of the differential metabolites are shown in Figure 4a. A hypothetical pathways investigation is shown in Figure 5.

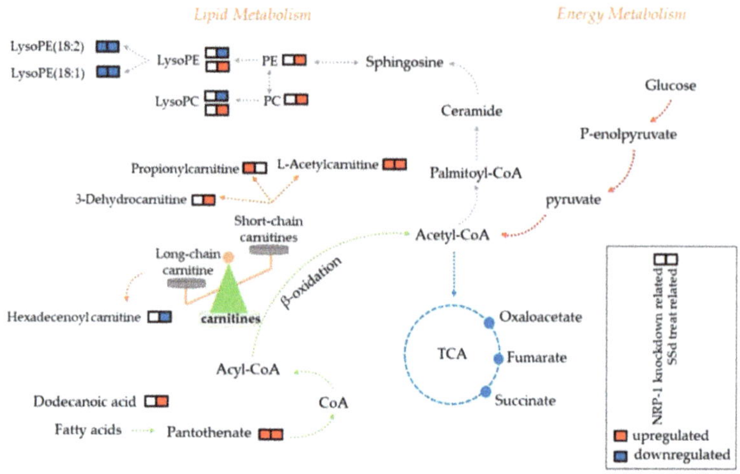

Figure 5. The possible perturbed metabolic pathways in response to NRP-1 knockdown and/or SSd. Upregulated (red) and downregulated (blue) levels of metabolites observed are indicated.

3. Discussion

In our study, we collected targets and investigated the affinity among SSd and potential proteins. Results showed that COX-2, NRP-1, and transcription factor p65 were the top three based on docking scores, which suggested the existence of strong interactions with SSd. Cancer-related targets such as COX-2, transcription factor p65, cellular tumor antigen p53 [31], and myc proto-oncogene protein [32] have been reported to be influenced by SSd. Other reported inflammatory and apoptosis-related targets associated with SSd also gained satisfactory docking scores, which proved the reliability of the methods applied. However, up to now no research has been reported on the relationship between NRP-1 and SSd.

In our research, NRP-1 was found to be upregulated on the 3.75 μM and 7.5 μM of SSd, while the cell viability of HepG2 was maintained at approximately 75%. Overexpression of NRP-1 was reported to promote the proliferation of cancer cells [20], which probably suggested that the cell damage effect of 7.5 μM SSd was antagonized by the proliferation promotion of upregulating NRP-1. This leads us to ask whether NRP-1 knockdown would enhance the damage effect of SSd at 7.5 μM, due to the upregulated NRP-1 and unexpected side effects appearing simultaneously with the anti-hepatoma effect of SSd. After knockdown of NRP-1, the results showed that, compared to the siNRP-1- and NC-treated SSd groups, the cell viability and wound healing percentage of siNRP-1 treated with SSd decreased significantly, not counting the damage of the transfection reagent. The outcomes above have clarified that NRP-1 knockdown enhanced the cell damage of SSd. NRP-1 acts as a co-receptor of VEGF, which has been proven essential for the progression and metastasis for tumors [39]. However, studies showed that adding external VEGF did not have an impact on the apoptosis of cancer cells when NRP-1 was knocked down, which proved the crucial role of NRP-1 in the anti-tumor effect of VEGF/NRP-1 [40]. Our study provided evidence of the enhanced anti-hepatoma effect of SSd on hepatocellular carcinoma cells with NRP-1 knockdown. However, such a hypothesis still remains to be confirmed in vivo.

To further explore the possible mechanism, a metabolomics study was performed. PCA and OPLS-DA proved the overall difference between groups. MDS was calculated as an indicator to quantitively evaluate the extent of disturbance by NRP-1 knockdown and SSd on HepG2 [41–43]. The results showed that NRP-1 knockdown exhibited the lowest MDS, which demonstrated that the global metabolic profile would not be severely influenced. However, SSd treated alone or NRP-1 knockdown in combination with SSd groups had a significant influence. These results suggested that SSd would have a great influence on the metabolism of both untreated and NRP-1 knockdown cells, in accordance with the more differential metabolites related to SSd. Even though it was not demonstrated by MDS, the elevated damage effect of NRP-1 knockdown in combination with SSd on HepG2 has been verified by our in vitro study.

In our study, four annotated differential metabolites related to both NRP-1 knockdown and SSd treatment were observed to have the same trend during an intervention, which suggested that the combined action of those metabolites contributed to the advanced anti-hepatoma effect. There were also differential metabolites influenced by SSd or NRP-1 knockdown, respectively. We propose that combining the action of common and particular metabolites contributed to the intensive effect. The annotated differential metabolites mainly included short- and long-chain acetylcarnitines, pantothenate, lyso-phosphatidylethanolamine (LPE), lyso-phosphatidylcholines (LPC), phosphatidylethanolamine (PE), and phosphatidylcholines (PC). Thus, we analyzed the possible underlying mechanisms.

Among all the annotated metabolites, pantothenate and acetylcarnitine were related to both NRP-1 knockdown and SSd, a short-chain carnitine, propionylcarnitine was elevated by NRP-1 knockdown, while an intermediate in carnitine degradation, 3-dehydrocarnitine, and a long-chain carnitine, hexadecenoyl carnitine, were detected to be influenced by SSd. A short-chain carnitine, acetylcarnitine, increased in our study; however, the long-chain carnitine was decreased significantly. The elevation of 3-dehydrocarnitine was probably related to the degradation of long-chain carnitines [44]. Both short- and long-chain acetylcarnitines are known to play important roles in fatty acid metabolism. Long-chain acetylcarnitines are responsible for transporting long-chain fatty acids into mitochondria. Fatty acids are then oxidized and converted to energy by the TCA cycle [45,46]. The decrease in long-chain carnitines suggested a blockage of the energy production in cancer cells. For the excess short- and medium-chain fatty acids, acetylcarnitine was responsible for the removal from the mitochondria. The increased acetylcarnitine implied the accumulation of short- and medium-chain fatty acids. Others also reported that increased long-chain carnitine and decreased short-chain carnitine were observed in hepatocellular carcinoma [47]. Combined in our study, the increase of short-chain carnitines and decrease of the long-chain carnitine probably contribute to the enhanced damage to HepG2. However, such a hypothesis should be verified by further research.

Lyso-phosphatidylethanolamine (LPE), lyso-phosphatidylcholines (LPC), phosphatidylethanolamine (PE), and phosphatidylcholines (PC) were influenced by SSd treatment, while only LPE (18:1) and LPE (18:2) were observed to be downregulated by both NRP-1 knockdown and SSd treatment. Four LPEs decreased and one increased under the intervention of SSd. Three LPCs increased and two decreased after treatment with SSd, whereas all the PEs and PCs were increased. Those reflected the dysregulation in phospholipid metabolism. LPC is derived from PC, which is a component of cell membranes and mitochondrial membrane. Three methylation reactions can transform PE to PC [48]. The results showed that a change of LPE (18:1) and LPE (18:2) possibly contributed to the cell damage caused by NRP-1 knockdown and SSd, while disorder in phospholipid metabolism was mainly induced by SSd rather than NRP-1 knockdown. Out metabolomic studies indicated that the enhanced anti-hepatoma effect of NRP-1 knockdown, in combination with SSd, mainly influenced the lipid metabolism, but this should be validated in the future.

4. Materials and Methods

4.1. Chemicals and Reagents

Cell culture medium RPMI 1640, DMEM, penicillin, and streptomycin were purchased from Gibco (Carlsbad, CA, USA). Saikosaponin d (purity > 98%) was purchased from Sichuan Victory Biological Technology Co., Ltd. (Chengdu, China). Fetal bovine serum (FBS) was purchased from Biological Industries (Kibbutz Beit-Haemek, Israel). Cell Counting Kit-8 detection kit and BCA protein assay kit were obtained from Beyotime (Shanghai, China). PVDF membrane and chemiluminescence (ECL) reagent were purchased from Millipore (Burlington, MA, USA). The primary antibodies were obtained from Cell Signaling Technology (Boston, MA, USA) and HRP-conjugated secondary antibodies were purchased from Proteintech (Philadelphia, PA, USA). The siNRP-1 sequences were provided and synthesized by GenePharma (Shanghai, China). Lipofectamine 2000 reagent was purchased from Invitrogen (Carlsbad, CA, USA). Heptadecanoic acid, glibenclamide, methoxyamine hydrochloride, *N*-methyl-*N*-trifluoroacetamide (MSTFA) and pyridine were purchased from Sigma-Aldrich (St. Louis, MO, USA). Methanol and ethyl acetate (HPLC grade) were purchased from Merck (Darmstadt, Germany). Distilled water was purified by a Milli-Q system (Merckmillipore, Darmstadt, Germany).

4.2. Target Collection and Screening

4.2.1. Target Collection

First, we searched several databases for the included SSd targets such as Traditional Chinese Medicine Systems Pharmacology Database and Analysis Platform (TCMSP version 2.3, http://lsp.nwu.edu.cn/tcmsp.php), Herbal Ingredients' Targets Database (HIT, http://lifecenter.sgst.cn/hit/) and Traditional Chinese Medicine Integrative Database (TCMID, https://academic.oup.com/nar/article/41/D1/D1089/1057998). The keywords for database searching were "saikosaponin d."

Secondly, reported targets of SSd were collected from literature mining. The keywords for literature mining were "saikosaponin d" and "targets."

Lastly, we predicted targets by chemical similarity via the Pubchem resource (https://pubchem.ncbi.nlm.nih.gov/). Search entries of three databases, BioAssay, Compound, and Substance, can be reached on the homepage. Our protocols were in three main steps, as follows: (1) Structure download. The 2D structure formula of SSd was obtained via the Compound entry. (2) Selecting bioactive compounds. After uploading the *sdf* format of SSd in Identity/Similarity searching, similar scores were defined by 95%. Then, we filtered bioactive compounds of SSd and its structure analogs that have been identified via BioAssays, Active. (3) Tracking down potential targets. Targets of bioactive compounds were listed in Bioactivity Analysis. For all the targets, only those with more than one corresponding active compound can be subsequently collected as putative targets. Targets of SSd and its structural analogs could be obtained through this method.

For the targets achieved above, unified standard protein names and gene names were searched in the Uniprot (https://www.uniprot.org/) database. Then the target numbers from the three sources were input into Venny 2.1.0 (http://bioinfogp.cnb.csic.es/tools/venny/index.html) to view overlapping ones and test the reliability of the methods.

4.2.2. Target Screening by Molecular Docking

To investigate the binding affinity of SSd and the candidate targets, molecular docking was carried out by Maestro 9.6 (Schrödinger, Inc, San Diego, CA, USA). First, the crystal structures of candidate proteins with high resolution and small ligands contained in the structures were downloaded from RCSB PDB (https://www.rcsb.org/). Next, we preprocessed the proteins by deleting the water molecules and other ligands. After energy minimizing, receptor-grid files and active sites were generated. Finally, SSd was docked at the active sites of the proteins after hydrogenating and charging. The targets were then sorted by docking scores and the degree of novelty.

4.3. Verification of the Target on HepG2

4.3.1. Cell Culture

Human hepatocellular carcinoma cell line HepG2 was obtained from Type Culture Collection Cell Bank (Chinese Academy of Sciences Committee, Shanghai, China), human hepatocellular carcinoma cell lines SMMC-7721 and human hepatocyte L-02 were kindly provided by Prof. Yong Yang (Center for New Drug Safety Evaluation and Research, China Pharmaceutical University). SMMC-7721 and

L-02 cell lines were cultured in RPMI 1640 with 10% FBS, 100 U/mL penicillin and 100 mg/mL streptomycin. HepG2 were cultured in DMEM with 10% FBS, 100 U/mL penicillin and 100 mg/mL streptomycin. All cells were cultured at 37 °C in a humidified incubator (Thermo, Waltham, MA, USA) with 5% CO_2.

4.3.2. Cell Proliferation Assay

The viability of HepG2 was measured by a Cell Counting Kit-8 (CCK-8) detection kit after treatment with different concentrations (1.75, 3.75, 7.5, 15 μM) of SSd. Briefly, 5×10^3 cells per well were seeded in 96-well plates and treated with SSd. After 24 h treatment, CCK-8 was added and absorbance at 450 nm was measured after 2 h incubation at 37 °C.

4.3.3. Cell Migration Assay

Migration of HepG2 was determined by a wound healing assay. The scratches were made after cell adherence in six-well plates. After rinsing with PBS, a medium containing different concentrations (1.75, 3.75, 7.5, 15 μM) of SSd and no FBS was replaced. Pictures obtained at 0 h and 24 h after scratching and distance of migration were analyzed by Image J (Version 1.48, National Institutes of Health, Bethesda, MD, USA).

4.3.4. Western Blot

For the Western blot analysis of NRP-1 expression in blank HepG2, SMMC-7721 and L-02, protein levels of the three cell lines were determined by a BCA protein assay kit. Proteins were also measured for the Western blot analysis of NRP-1 expression in HepG2 after treatment with different concentrations of SSd. Thirty micrograms of protein per well were resolved by SDS-polyacrylamide gel, transferred to PVDF membrane, and blocked with 5% skim milk in phosphate-buffered saline containing 0.1% Tween 20. The membranes were incubated with primary antibodies at 4 °C overnight and probed with HRP-conjugated secondary antibodies at room temperature on the following day. The bands were visualized by Tanon (Shanghai, China) after using a chemiluminescence (ECL) reagent.

4.3.5. Knockdown of NRP-1 Using Small Interfering RNA

HepG2 were cultured in a six-well plate then transfected with a mixture of Lipofectamine 2000 reagent and siRNAs in serum-free DMEM for 6 h. Cells were collected and we detected the knockdown efficiency by Western blot after incubating them for 48 h.

4.3.6. Effect of SSd on siNRP-1 Cells

After transfection with siNRP-1/NC on HepG2 for 48 h, the medium was discarded and 7.5 μM SSd was treated for another 24 h. Cell viability and migration were then measured to evaluate the effect of NRP-1 knockdown in combination of SSd.

4.3.7. Statistical Analysis

Data analyses were performed by SPSS 19.0 software (Chicago, IL, USA) and results were expressed as mean ± standard deviation (SD). The statistical significance of differences between the two groups was indicated by a Student's t test. $p < 0.05$ was considered statistically significant.

4.4. Untargeted Metabolomics Analysis

4.4.1. Sample Collecting and Metabolomic Analysis

HepG2 were seeded in culture plates and divided into four groups (siNRP-1 group, NC group, siNRP-1 and SSd-treated group, NC and SSd-treated group, $n = 6$). Cells were treated as described above. Culture medium containing SSd was removed before 3 mL cold methanol was added. Internal standard was dissolved in methanol firstly, 5 μg/mL heptadecanoic acid for GC, and 20 μg/mL glibenclamide for LC. Then the samples were quenched at −80 °C for 20 min. Cells were scraped, ultrasonicated for 3 min, and centrifuged for 10 min to collect the supernatant for subsequent procedures. We dried the extracts with nitrogen flow at 37 °C and redissolved them with methanol before metabolomic analysis. Sample preparation and methods of instruments were based on our previous studies [49–51], as shown in the Supplementary Materials.

GC-MS analysis was performed on Shimadzu GCMS-QP2010 Ultra (Ultra GC-Q/MS; Shimadzu Inc., Kyoto, Japan) equipped with an Rtx-5MS capillary column (30.0 m × 0.25 mm ID, 0.25 μm). LC-MS analysis was carried out on a Shimadzu Prominence series ultrafast liquid chromatography (UFLC) system coupled with an ion trap time-of flight mass spectrometry system (IT-TOF/MS) (Shimadzu Inc.). Separation was achieved by a Phenomenex Kinetex C18 column (100 mm × 2.1 mm, 2.6 μm) (Phenomenex, Torrance, CA, USA).

4.4.2. Data Preprocessing and Analysis

Data preprocessing and analysis were also based on our previous studies [52,53]. The original chromatogram was processed for peak deconvolution and alignment by Profiling Solution (Shimadzu, Kyoto, Japan, version 1.1). The primary parameters were set as follows: ion *m*/*z* tolerance (500 mDa for GC/MS and 20 mDa for LC/MS), ion retention time tolerance (0.1 min for GC/MS and 0.2 min for LC/MS), and ion intensity threshold (10,000 counts for GC/MS and 8000 counts for LC/MS). The data were then exported to Excel and handled according to the "80%" rule: only metabolites detected in at least 80% of one group or more would be kept. Variables with relative standard deviation (RSD) lower than 30% in quality control (QC) samples were retained for further analysis.

The preprocessed data were imported into SIMCA-P (Umetrics, Sweden, Version 13.0). After pareto scaling of all the variables, principal components analysis (PCA) and orthogonal partial least squares discriminant analysis (OPLS-DA) were performed. The normalization by total ion intensity was performed to standardize multivariant analysis. Differences among groups could be observed. The differential metabolites were selected by variable importance in the projection (VIP) and *p*-values obtained by Mann-Whitney U test, variables with VIP > 1 and $p < 0.05$ were screened out.

4.4.3. Quantitative Evaluation of NRP-1 Knockdown and SSd on HepG2 by Metabolites Deregulation Score (MDS)

Effects of NRP-1 knockdown and SSd treatment could be quantified by the metabolites deregulation score (MDS). All the metabolites detected were integrated as a whole dataset. Then the R package pathifier was applied to analyze the metabolite sets. The whole dataset was considered as a cloud and we built a "principal curve" through the NC samples in the cloud [44]. Since untargeted metabolomic method covers limited metabolic pathways, we calculated the overall MDS instead of the pathway deregulation score (PDS) by the R package pathifier [45,46]. Each sample was projected onto the principal curve and the projection distance was MDS.

4.4.4. Differential Metabolites Annotation

Annotation of metabolites detected by GC/MS was carried out by comparison with the National Institute of Standard and Technology library. Peaks with more than 80% similarity were assigned corresponding compound names. For LC/MS, mass-to-charge (*m*/*z*) and MS/MS fragmentation patterns of metabolites were compared with the information provided in the HMDB database (http://www.hmdb.ca/) for annotation. Then, retention time, accurate *m*/*z*, and the MS/MS fragmentation of features of interest were compared with those standard compounds available in our lab. Pathways by which metabolites got involved were searched and enriched on the MetaboAnalyst (version 4.0) and KEGG databases (https://www.kegg.jp/).

For the differential metabolites, NRP-1 knockdown and SSd-related were filtered to calculate the fold change of concentrations.

5. Conclusions

In our study, potential targets of SSd were predicted and NRP-1 was chosen for further research. Experimental proof of the enhanced anti-hepatoma effect of SSd in combination with NRP-1 knockdown was provided for the first time. Further metabolomics study revealed that lipid transportation and phospholipid metabolism were significantly altered when NRP-1 knockdown and SSd were treated on HepG2. Our findings indicate a new insight into the understanding of mechanism of SSd, but this should be confirmed in the future.

Supplementary Materials: The following are available online: Figure S1: The intersection of targets amount obtained from three sources; Figure S2: The pictures of cell migration assay after NRP-1 knockdown and/or SSd treatment; Table S1: Potential targets of SSd searched by databases; Table S2: Targets obtained by literature mining; Table S3: Targets and its active compounds predicted by SSd and its structural analogues in Pubchem; Table S4: All the predicted targets of SSd; Table S5: Statistical parameters of the models; Table S6: Differential metabolites relevant to both NRP-1 knockdown and SSd treatment; Table S7: Differential metabolites related to NRP-1 knockdown; Table S8: Differential metabolites related to SSd treated.

Author Contributions: Conceptualization, Y.L. and F.X.; Data curation, Y.L.; Funding acquisition, F.X.; Investigation, Y.L.; Methodology, Y.L.; Project administration, F.X.; Resources, F.X.; Software, Y.C.; Supervision, F.X.; Validation, X.H. and Q.Z.; Visualization, R.L., L.X., Y.T., R.S., Z.Z. and F.X.; Writing—original draft, Y.L.

Funding: This research received no external funding.

Acknowledgments: This work was supported by the NSFC (No. 81573626), Jiangsu Provincial National Science Foundation for Distinguished Young Scholars (No. BK20180027), Double First-Class University project, the Program for Jiangsu province Innovative Research Team, and a project funded by the Priority Academic Program Development of Jiangsu Higher Education Institutions (PAPD).

Conflicts of Interest: The authors declare no conflict of interest.

References

1. Dang, S.S.; Wang, B.F.; Cheng, Y.A.; Song, P.; Liu, Z.G.; Li, Z.F. Inhibitory effects of saikosaponin-d on CCl4-induced hepatic fbrogenesis in rats. *World J. Gastroenterol.* **2007**, *13*, 557–563. [CrossRef]
2. Lu, C.N.; Yuan, Z.G.; Zhang, X.L.; Yan, R.; Zhao, Y.Q.; Liao, M.; Chen, J.X. Saikosaponin a and its epimer saikosaponin d exhibit anti-inflammatory activity by suppressing activation of NF-kappaB signaling pathway. *Int. Immunopharmacol.* **2012**, *14*, 121–126. [CrossRef]
3. Wang, B.F.; Dai, Z.J.; Wang, X.J.; Bai, M.H.; Lin, S.; Ma, H.B.; Wang, Y.L.; Song, L.Q.; Ma, X.L.; Zan, Y.; et al. Saikosaponin-d increases the radiosensitivity of smmc-7721 hepatocellular carcinoma cells by adjusting the g0/g1 and g2/m checkpoints of the cell cycle. *BMC Complement. Altern. Med.* **2013**, *13*. [CrossRef]
4. Wong, V.K.; Zhou, H.; Cheung, S.S.; Li, T.; Liu, L. Mechanistic study of saikosaponin-d (Ssd) on suppression of murine T lymphocyte activation. *J. Cell. Biochem.* **2009**, *107*, 303–315. [CrossRef]
5. Li, Z.Y.; Jiang, Y.M.; Liu, Y.M.; Guo, Z.; Shen, S.N.; Liu, X.M.; Pan, R.L. Saikosaponin D acts against corticosterone-induced apoptosis via regulation of mitochondrial GR translocation and a GR-dependent pathway. *Prog. Neuropsychopharmacol. Biol. Psychiatry* **2014**, *53*, 80–89. [CrossRef]
6. Hao, Y.; Piao, X.; Piao, X. Saikosaponin-d inhibits beta-conglycinin induced activation of rat basophilic leukemia-2H3 cells. *Int. Immunopharmacol.* **2012**, *13*, 257–263. [CrossRef]
7. Zhang, F.; Chen, L.; Jin, H.; Shao, J.; Wu, L.; Lu, Y.; Zheng, S. Activation of Fas death receptor pathway and Bid in hepatocytes is involved in saikosaponin D induction of hepatotoxicity. *Environ. Toxicol. Pharmacol.* **2016**, *41*, 8–13. [CrossRef]
8. Chen, M.F.; Huang, S.J.; Huang, C.C.; Liu, P.S.; Lin, K.I.; Liu, C.W.; Hsieh, W.C.; Shiu, L.Y.; Chen, C.H. Saikosaponin d induces cell death through caspase-3-dependent, caspase-3-independent and mitochondrial pathways in mammalian hepatic stellate cells. *BMC Cancer* **2016**, *16*, 532. [CrossRef]
9. Chen, L.; Zhang, F.; Kong, D.; Zhu, X.; Chen, W.; Wang, A.; Zheng, S. Saikosaponin D disrupts platelet-derived growth factor-beta receptor/p38 pathway leading to mitochondrial apoptosis in human LO2 hepatocyte cells: A potential mechanism of hepatotoxicity. *Chem. Biol. Interact.* **2013**, *206*, 76–82. [CrossRef]
10. Ye, H.; Ye, L.; Kang, H.; Zhang, D.; Tao, L.; Tang, K.; Liu, X.; Zhu, R.; Liu, Q.; Chen, Y.Z.; et al. HIT: Linking herbal active ingredients to targets. *Nucleic Acids Res.* **2011**, *39*, D1055–D1059. [CrossRef]

11. Huang, L.; Xie, D.; Yu, Y.; Liu, H.; Shi, Y.; Shi, T.; Wen, C. TCMID 2.0: A comprehensive resource for TCM. *Nucleic Acids Res.* **2018**, *46*, D1117–D1120. [CrossRef]
12. Ru, J.; Li, P.; Wang, J.; Zhou, W.; Li, B.; Huang, C.; Li, P.; Guo, Z.; Tao, W.; Yang, Y.; et al. TCMSP: A database of systems pharmacology for drug discovery from herbal medicines. *J. Cheminform.* **2014**, *6*, 13. [CrossRef]
13. Mathias, D.; Stefan, G.; Jessica, A.; Burghardt, W.; Robert, P. SuperPred: Drug classification and target prediction. *Nucleic Acids Res.* **2008**, *36*, 55–59. [CrossRef]
14. Nikolai, H.; Jessica, A.; Joachim, V.E.; Mathias, D.; Karel, M.; Andreas, E.; Gilson, M.K.; Bourne, P.E.; Robert, P. SuperTarget goes quantitative: Update on drug-target interactions. *Nucleic Acids Res.* **2012**, *40*, 1113–1117. [CrossRef]
15. Liu, T.; Lin, Y.; Wen, X.; Jorissen, R.N.; Gilson, M.K. BindingDB: A web-accessible database of experimentally determined protein-ligand binding affinities. *Nucleic Acids Res.* **2007**, *35*, D198–D201. [CrossRef]
16. Rognan, D. Chemogenomic approaches to rational drug design. *Br. J. Pharmacol.* **2007**, *152*, 38–52. [CrossRef]
17. Pellet-Many, C.; Frankel, P.; Jia, H.; Zachary, I. Neuropilins: Structure, function and role in disease. *Biochem. J.* **2008**, *411*, 211–226. [CrossRef]
18. Zachary, I.C.; Frankel, P.; Evans, I.M.; Pellet-Many, C. The role of neuropilins in cell signalling. *Biochem. Soc. Trans.* **2009**, *37*, 1171–1178. [CrossRef]
19. Fuh, G.; Garcia, K.C.; de Vos, A.M. The interaction of neuropilin-1 with vascular endothelial growth factor and its receptor flt-1. *J. Biol. Chem.* **2000**, *275*, 26690–26695. [CrossRef]
20. Berge, M.; Allanic, D.; Bonnin, P.; de Montrion, C.; Richard, J.; Suc, M.; Boivin, J.F.; Contreres, J.O.; Lockhart, B.P.; Pocard, M.; et al. Neuropilin-1 is upregulated in hepatocellular carcinoma and contributes to tumour growth and vascular remodelling. *J. Hepatol.* **2011**, *55*, 866–875. [CrossRef]
21. Tse, B.W.C.; Volpert, M.; Ratther, E.; Stylianou, N.; Nouri, M.; McGowan, K.; Lehman, M.L.; McPherson, S.J.; Roshan-Moniri, M.; Butler, M.S.; et al. Neuropilin-1 is upregulated in the adaptive response of prostate tumors to androgen-targeted therapies and is prognostic of metastatic progression and patient mortality. *Oncogene* **2017**, *36*, 3417–3427. [CrossRef]
22. Li, L.; Jiang, X.; Zhang, Q.; Dong, X.; Gao, Y.; He, Y.; Qiao, H.; Xie, F.; Xie, X.; Sun, X. Neuropilin-1 is associated with clinicopathology of gastric cancer and contributes to cell proliferation and migration as multifunctional co-receptors. *J. Exp. Clin. Cancer Res.* **2016**, *35*, 16. [CrossRef]
23. Jubb, A.M.; Strickland, L.A.; Liu, S.D.; Mak, J.; Schmidt, M.; Koeppen, H. Neuropilin-1 expression in cancer and development. *J. Pathol.* **2012**, *226*, 50–60. [CrossRef]
24. Casazza, A.; Laoui, D.; Wenes, M.; Rizzolio, S.; Bassani, N.; Mambretti, M.; Deschoemaeker, S.; Van Ginderachter, J.A.; Tamagnone, L.; Mazzone, M. Impeding macrophage entry into hypoxic tumor areas by Sema3A/Nrp1 signaling blockade inhibits angiogenesis and restores antitumor immunity. *Cancer Cell* **2013**, *24*, 695–709. [CrossRef]
25. Tordjman, R.; Lepelletier, Y.; Lemarchandel, V.; Cambot, M.; Gaulard, P.; Hermine, O.; Romeo, P.H. A neuronal receptor, neuropilin-1, is essential for the initiation of the primary immune response. *Nat. Immunol.* **2002**, *3*, 477–482. [CrossRef]
26. Nicholson, J.K.; Connelly, J.; Lindon, J.C.; Holmes, E. Metabonomics: A platform for studying drug toxicity and gene function. *Nat. Rev. Drug Discov.* **2002**, *1*, 153–161. [CrossRef]
27. Li, Q.; Cheng, T.; Wang, Y.; Bryant, S.H. PubChem as a public resource for drug discovery. *Drug Discov. Today* **2010**, *15*, 1052–1057. [CrossRef]
28. Lu, X.L.; He, S.X.; Ren, M.D.; Wang, Y.L.; Zhang, Y.X.; Liu, E.Q. Chemopreventive effect of saikosaponin-d on diethylinitrosamine-induced hepatocarcinogenesis: Involvement of CCAAT/enhancer binding protein β and cyclooxygenase-2. *Mol. Med. Rep.* **2012**, *5*, 637–644. [CrossRef]
29. Liu, A.; Tanaka, N.; Sun, L.; Guo, B.; Kim, J.H.; Krausz, K.W.; Fang, Z.; Jiang, C.; Yang, J.; Gonzalez, F.J. Saikosaponin d protects against acetaminophen-induced hepatotoxicity by inhibiting NF-κB and STAT3 signaling. *Chem. Biol. Int.* **2014**, *223*, 80–86. [CrossRef]
30. Zu, N.; Li, P.; Li, N.; Choy, P.; Gong, Y. Mechanism of saikosaponin-d in the regulation of rat mesangial cell proliferation and synthesis of extracellular matrix proteins. *Biochem. Cell Biol.* **2007**, *85*, 169–174. [CrossRef]
31. Hsu, Y.L.; Kuo, P.L.; Lin, C.C. The proliferative inhibition and apoptotic mechanism of Saikosaponin D in human non-small cell lung cancer A549 cells. *Life Sci.* **2004**, *75*, 1231–1242. [CrossRef]

32. Lin, X.; Wu, S.; Wang, Q.; Shi, Y.; Liu, G.; Zhi, J.; Wang, F. Saikosaponin-D Reduces H_2O_2-Induced PC12 Cell Apoptosis by Removing ROS and Blocking MAPK-Dependent Oxidative Damage. *Cell. Mol. Neurobiol.* **2016**, *36*, 1–11. [CrossRef]
33. Fan, J.; Li, X.; Li, P.; Li, N.; Wang, T.; Shen, H.; Siow, Y.; Choy, P.; Gong, Y. Saikosaponin-d attenuates the development of liver fibrosis by preventing hepatocyte injury. *Biochem. Cell Biol.* **2007**, *85*, 189–195. [CrossRef]
34. Chiang, L.C.; Ng LTLiu, L.T.; Shieh, D.E.; Lin, C.C. Cytotoxicity and anti-hepatitis B virus activities of saikosaponins from bupleurum species. *Planta Med.* **2003**, *69*, 705–709. [CrossRef]
35. Hsu, M.J.; Cheng, J.S.; Huang, H.C. Effect of saikosaponin, a triterpene saponin, on apoptosis in lymphocytes: Association with c-myc, p53, and bcl-2 mRNA. *Br. J. Pharmacol.* **2000**, *131*, 1285–1293. [CrossRef]
36. Xu, J.; Xia, J. NRP-1 silencing suppresses hepatocellular carcinoma cell growth in vitro and in vivo. *Exp. Ther. Med.* **2013**, *5*, 150–154. [CrossRef]
37. Pan, Q.; Chanthery, Y.; Liang, W.C.; Stawicki, S.; Mak, J.; Rathore, N.; Tong, R.K.; Kowalski, J.; Yee, S.F.; Pacheco, G.; et al. Blocking neuropilin-1 function has an additive effect with anti-VEGF to inhibit tumor growth. *Cancer Cell* **2007**, *11*, 53–67. [CrossRef]
38. Zhang, Y.; Liu, P.; Jiang, Y.; Dou, X.; Yan, J.; Ma, C.; Fan, Q.; Wang, W.; Su, F.; Tang, H.; et al. High Expression of Neuropilin-1 Associates with Unfavorable Clinicopathological Features in Hepatocellular Carcinoma. *Pathol. Oncol. Res.* **2016**, *22*, 367–375. [CrossRef]
39. Ferrara, N.; Gerber, H.P.; LeCouter, J. The biology of VEGF and its receptors. *Nat. Med.* **2003**, *9*, 669–676. [CrossRef]
40. Luo, M.; Hou, L.; Li, J.; Shao, S.; Huang, S.; Meng, D.; Liu, L.; Feng, L.; Xia, P.; Qin, T.; et al. VEGF/NRP-1axis promotes progression of breast cancer via enhancement of epithelial-mesenchymal transition and activation of NF-kappaB and beta-catenin. *Cancer Lett.* **2016**, *373*, 1–11. [CrossRef]
41. Cui, D.N.; Wang, X.; Chen, J.Q.; Lv, B.; Zhang, P.; Zhang, W.; Zhang, Z.J.; Xu, F.G. Quantitative Evaluation of the Compatibility Effects of Huangqin Decoction on the Treatment of Irinotecan-Induced Gastrointestinal Toxicity Using Untargeted Metabolomics. *Front. Pharmacol.* **2017**, *8*, 211. [CrossRef]
42. Drier, Y.; Sheffer, M.; Domany, E. Pathway-based personalized analysis of cancer. *Proc. Natl. Acad. Sci. USA* **2013**, *110*, 6388–6393. [CrossRef]
43. Huang, S.; Chong, N.; Lewis, N.E.; Jia, W.; Xie, G.; Garmire, L.X. Novel personalized pathway-based metabolomics models reveal key metabolic pathways for breast cancer diagnosis. *Genome Med.* **2016**, *8*, 34. [CrossRef]
44. Liang, Q.; Liu, H.; Zhang, T.; Jiang, Y.; Xing, H.; Zhang, A.-h. Metabolomics-based screening of salivary biomarkers for early diagnosis of Alzheimer's disease. *RSC Adv.* **2015**, *5*, 96074–96079. [CrossRef]
45. Hoppel, C. The role of carnitine in normal and altered fatty acid metabolism. *Am. J. Kidney Dis.* **2003**, *41*, S4–S12. [CrossRef]
46. Zhou, L.; Wang, Q.; Yin, P.; Xing, W.; Wu, Z.; Chen, S.; Lu, X.; Zhang, Y.; Lin, X.; Xu, G. Serum metabolomics reveals the deregulation of fatty acids metabolism in hepatocellular carcinoma and chronic liver diseases. *Anal. Bioanal. Chem.* **2012**, *403*, 203–213. [CrossRef]
47. Lu, Y.; Li, N.; Gao, L.; Xu, Y.J.; Huang, C.; Yu, K.; Ling, Q.; Cheng, Q.; Chen, S.; Zhu, M.; et al. Acetylcarnitine Is a Candidate Diagnostic and Prognostic Biomarker of Hepatocellular Carcinoma. *Cancer Res.* **2016**, *76*, 2912–2920. [CrossRef]
48. Ren, C.; Liu, J.; Zhou, J.; Liang, H.; Wang, Y.; Sun, Y.; Ma, B.; Yin, Y. Lipidomic analysis of serum samples from migraine patients. *Lipids Health Dis.* **2018**, *17*, 22. [CrossRef]
49. Guo, Q.; Zhang, Q.-Q.; Chen, J.-Q.; Zhang, W.; Qiu, H.-C.; Zhang, Z.-J.; Liu, B.-M.; Xu, F.-G. Liver metabolomics study reveals protective function of Phyllanthus urinaria against CCl 4 -induced liver injury. *Chin. J. Nat. Med.* **2017**, *15*, 525–533. [CrossRef]
50. Wang, X.; Cui, D.N.; Dai, X.M.; Wang, J.; Zhang, W.; Zhang, Z.J.; Xu, F.G. HuangQin Decoction Attenuates CPT-11-Induced Gastrointestinal Toxicity by Regulating Bile Acids Metabolism Homeostasis. *Front. Pharmacol.* **2017**, *8*, 156. [CrossRef]
51. Zhang, M.-H.; Chen, J.-Q.; Guo, H.-M.; Li, R.-T.; Gao, Y.-Q.; Tian, Y.; Zhang, Z.-J.; Huang, Y. Combination of LC/MS and GC/MS based metabolomics to study the hepatotoxic effect of realgar nanoparticles in rats. *Chin. J. Nat. Med.* **2017**, *15*, 684–694. [CrossRef]

52. Wong, V.K.W.; Molly Miao, Z.; Hua, Z.; Lam, K.Y.C.; Po Ling, C.; Law, C.K.M.; Yue, P.Y.K.; Liang, L. Saikosaponin-d Enhances the Anticancer Potency of TNF-α via Overcoming Its Undesirable Response of Activating NF-Kappa B Signalling in Cancer Cells. *Evid.-Based Complement. Altern. Med.* **2013**, 745295. [CrossRef]
53. He, S.; Lu, G.; Hou, H.; Zhao, Z.; Zhu, Z.; Lu, X.; Chen, J.; Wang, Z. Saikosaponin-d suppresses the expression of cyclooxygenase-2 through the phospho-signal transducer and activator of transcription 3/hypoxia-inducible factor-1α pathway in hepatocellular carcinoma cells. *Mol. Med. Rep.* **2014**, *10*, 2556. [CrossRef]

Sample Availability: Samples of the compounds are not available from the authors.

MDPI
St. Alban-Anlage 66
4052 Basel
Switzerland
Tel. +41 61 683 77 34
Fax +41 61 302 89 18
www.mdpi.com

Molecules Editorial Office
E-mail: molecules@mdpi.com
www.mdpi.com/journal/molecules